Visual Fitness

Visual Fitness

7 Minutes to Better Eyesight and Beyond

David Cook, O.D.

BERKLEY BOOKS, NEW YORK

A Berkley Book
Published by The Berkley Publishing Group
A division of Penguin Group (USA) Inc.
375 Hudson Street,
New York, New York 10014

This book is an original publication of The Berkley Publishing Group

Cover design by Jill Boltin.
Cover photograph by Siede Preis/Photo Disc.
Text design by Tiffany Estreicher

PRINTING HISTORY
Berkley trade paperback edition / February 2004
Berkley trade paperback ISBN: 0-425-19408-6

Library of Congress Cataloging-in-Publication Data
Cook, David L.
Visual fitness : seven minutes to better eyesight and beyond / David Cook.
p. cm.
Includes bibliographical references.
ISBN 0-425-19408-6
1. Eye—Care and hygiene—Popular works. 2. Eye—Diseases—Prevention—Popular works. I. Title.

RE51.C76 2004
617.7—dc22

2003065921

PRINTED IN THE UNITED STATES OF AMERICA

10 9 8 7 6 5 4 3 2 1

To the memory of Dr. William M. Ludlam,
teacher, mentor, colleague, friend

Acknowledgments

Thank you Jane Dystel and Michael Bourret for being great agents. Thank you Denise Silvestro, Martha Bushko, and Sheila Moody for being great editors. Thank you Shannon Sharpe and Deborah Kohan for being great publicists. Thank you Vickki Ford Cook for being a great wife and Anya Ford Cook for being potty trained before joining us from Russia—certainly a sign of greatness for an adopted three year old.

I'd also like to thank the following doctors of optometry for their insights, support, and friendship along the way: Joyce Adema, Stanley Appelbaum, Beth Ballinger, Curtis Baxstrom, Beth Bazin, Ira Bernstein, Ron Bateman, Martin Birnbaum, James Bosse, Carol Burns, Carl Childress, Mary Childress, Garth Christenson, Kathleen Clark, Kevin Cline, Allen Cohen, Barry Cohen, Stanford Cohen, Jeffrey Cooper, Ron Danner, Roger Dowis, Lou Spinozzi, Mitchell Davis, Nicholas Depotidis, Neil Draisin, Michael Egger, Stanley Evans, Gary Etting, Rod Fields, Daniel Fleming, Elliot Forrest, Dan Fortenbacher, Robert Fox, Michael Frier, Melvin Fox, Michael Gallaway, Don Getz, Drusilla Grant, Nelson Greeman, Israel Greenwald, Anthony Grieneisen, Sid Groffman, Carl Grunning, Steven Haleo, Paul Harris, Marilyn Heinke, Lynn Hellerstein, Carl Hillier, Robert Hohendorf, Sam Horner Junior, Sam Horner III, Donald Janiuk, Colin Kageyama, Garry Kapel, Richard Kavner, Jeffrey Keene, Gregory Kitchener, Robert Kraskin, Lawrence Lampert, William Leadingham, Paul Lederer, Stanley Levine, Robin Lewis, Sue Lowe, William Ludlam, W. C. Maples, Carol Marusich, James Mayer, Michael McQuillan, Paul McRae, Richard Meier,

Joseph Miele, Brenda Montecalvo, Lori Mowbray, William Moskowitz, Steven Nicholson, Chris Peterson, Michael Politzer, Paul Rousseau, Leonard Press, Frank Puckett, Kristy Remick, Larry Ritter, Marcy Rose, Eldon Rosenow, Austin Ruiz, Stuart Rothman, Robert Sanet, Robin Sapossnek, Mitchell Scheiman, Carol Scott, Arthur Seiderman, Diane Serex-Dougan, Steven Shaby, Kathy Shamblin, Denise Shepard, Arnold Sherman, Bradford Smith, Harold Solan, Glen Steele, John Streff, Irwin Suchoff, Barry Tannen, Elizabeth Todd, Robert Toler, Nancy Torgerson, Claude Valenti, Mary Van Hoy, Gary Veronneau, Harry Wachs, Catherine Ward, Daniel Weinberg, Gary Williams, Bruce Wojciechowski, Mark Wright, Todd Wylie, Stanley Yamane, and Wayne Yorkgitis.

Contents

Introduction **xi**

1 What Is Visual Fitness and How May It Be Affecting Your Life? **1**

2 Your First Step: The Visual Fitness Self-Assessment **20**

3 **VISUAL FITNESS SKILL I: CLEAR-SIGHT**
How "Little" Can You See? **33**

4 **VISUAL FITNESS SKILL II: BIG-SIGHT**
How "Big" Can You See? **49**

5 **VISUAL FITNESS SKILL III: DEEP-SIGHT**
How Much Depth Can You See? **69**

6 **VISUAL FITNESS SKILL IV: STRONG-SIGHT**
How Long Can You See? **76**

7 **VISUAL FITNESS SKILL V: QUICK-SIGHT**
How Fast Can You See? 101

8 **VISUAL FITNESS SKILL VI: SMART-SIGHT**
Seeing with the Mind's Eye 156

9 **VISUAL FITNESS SKILL VII: SPORTS-SIGHT**
Keeping Your Seeing on the Ball 176

10 Your Visual Fitness Workouts:
Organizing for Success 193

11 Visual Fitness Trainers:
Who Are They? Where Did They
Come From? Why Should You Visit One? 200

12 Breaking the Sight Barrier:
Seeing with the Power of Choice 210

APPENDIX A
Finding a Visual Fitness Coach 215

APPENDIX B
Research on Vision Therapy 217

Bibliography 221

Introduction

As infants, we entered the world legally blind. We could distinguish little more than vague, watery images, which, for most of us, later focused into the love on a parent's face.

As we developed, we followed the path mapped out by our genes and learned to see. That seeing varies. Take President John Kennedy, for instance; he read a thousand words per minute. Consider baseball legend Ted Williams; he tracked the seams of a ball at one hundred miles per hour. These are examples of what seeing can, and should, be. Such seeing—which reaches beyond simple eyesight—is not just the product of better genes or the tissues of the eyes themself. Such seeing is learned. It depends on how we *use* our eyes.

Few reach their seeing potential. No matter how well we read the eye chart in the doctor's office, some still have trouble focusing on the print in books for long periods, while others can't seem to see the big picture—that is, their peripheral vision goes unused. Some of us struggle to see at night; still others experience problems with depth perception. The list goes on. President Kennedy couldn't have hit a baseball like Ted Williams; Ted couldn't have read with the president's speed. No matter what our seeing strengths, there are weaknesses as well. But it doesn't have to be this way. There is something we can do about it. And I'm not talking about changing your glasses or throwing them away or having the latest laser surgery. Your path to better eyesight and beyond begins with a seven-minute workout.

For twenty-five years now, as a professionally trained and licensed eye doctor, I've used simple exercises to help individuals to see more, in less time, with less effort. I've guided thousands to sight that is clearer, bigger, deeper, stronger, quicker, and smarter. Now it's time I worked with you. Whatever the strength of your glasses, however perfect you believe your sight to be, in the following pages I'll guide you through the very "instruction manual" you need for your eyes.

To understand why using this manual is necessary, even if your eyes are already perfect, consider Vickki's story.

When I first met Vickki, she was in her thirties with short blond hair, a heart-shaped face, and soft green eyes. Those eyes had served her well. They'd held the attention of judges when she was crowned Miss Georgia World. They'd captivated audiences while, for seventeen years, she supported herself and her towheaded son singing in nightclubs. They'd even seen her through the endless volumes of print she read when she returned to school to study law. But now, unbeknownst to Vickki, those eyes were failing her.

Vickki was studying for her bar exam. As she spent hour after hour answering sample questions, her natural enthusiasm began to wane. She knew that passing would require her to complete sixty questions per hour, but try as she would, she was finding it impossible to complete even half that number. And it seemed the harder she worked, the worse it became. Her eyes began to burn and her head began to ache.

Rather than be discouraged by her increasing discomfort, Vickki was actually heartened. She wondered if a pair of glasses wouldn't solve her problem. Full of hope, she took a morning away from the books to visit her eye doctor. That afternoon she resumed her studies more discouraged than ever. The doctor's pronouncement: "Your eyes are perfect, but the way you're pushing yourself, no wonder you have eyestrain. You need to take more breaks."

For several days she tried the breaks, but even though her comfort increased, her test-taking speed remained painfully slow, and besides, she reasoned, there would be little time for breaks during her test. Headaches or no, she returned to her former study schedule and trudged painfully on, wondering if perhaps she wasn't bright enough to be a lawyer.

Despite this frustrating beginning, Vickki's story has a happy ending. She was referred to my office to build her visual fitness. Eager to succeed, she incorporated the necessary workouts into her study schedule, and soon her test-taking rate had doubled. She passed her bar exam on her first try and married her visual fitness coach. After ten years, the two of us are still gazing athletically into each other's eyes.

But what is visual fitness?

The pages that follow will answer this question in a way that may not only surprise you, but may explain things about your performance in life that have puzzled and perhaps even disturbed you in the past. Even if your eyes are perfect, unless they came with an instruction manual, this book is for you.

The first chapter offers a basic overview of visual fitness. The second provides a self-assessment, which will allow you to discover the many unseen ways in which visual fitness may be affecting your life even if you have perfect eyes or perfect glasses. Each of the next seven chapters presents a different visual fitness skill. All seven of these chapters begin with general information and then describe your visual fitness workout. The general information sections will explain how the visual skills relate to everyday life. The workouts will offer the actual exercises you need to improve those skills.

Your first time through these seven chapters, I would suggest you see the big picture rather than get lost in the details. Read the information section. Perform the first exercise of the accompanying workout to clarify what you've read. Skip the rest of the exercises and move on to the next chapter. Don't spend time studying each exercise your first time through the book. There'll be plenty of time for that when your workouts begin in earnest.

Chapter 10 will put it all together so that when you do go back and learn the other exercises you'll be able to organize your workouts for success. For those who would prefer to hire a visual fitness pro, chapter 11 will tell you what you need to know about them, and Appendix A will help you find one. Finally, chapter 12 will help you to appreciate more of the potential impact of visual fitness on not only your eyes, but your outlook.

This book doesn't take the place of a regular eye exam. Nor is this book another program to throw away your glasses (although what you learn in this book may mean the difference between success and failure in such programs). Quite simply, this book is an instruction manual for the use of your eyes. If you'd like your seeing to be clearer, bigger, deeper, stronger, quicker, and smarter, just turn the page and read the instructions.

1

What Is Visual Fitness and How May It Be Affecting Your Life?

A husband and wife visit the eye doctor. Their examinations reveal that the man needs glasses; the woman does not—her eyes are perfect. Which one of the couple has the best visual fitness?

To be honest, I can't tell you from the above information. No doctor can. Routine eye examinations are designed to reveal if you need glasses or if some hidden health problem is stealing away your sight. Such examinations are important. Ask anyone who has put on a first pair of glasses and discovered that there were leaves on the trees. Ask anyone who put off seeing the eye doctor until it was too late. Routine eye examinations can make the difference between sight and blindness. They do not, however, test for visual fitness.

So what is visual fitness?

The following analogy may help. Suppose you go to the doctor and have your hands examined. The examination uses the latest in technology to reveal that you have ten fingers, no missing knuckles, no arthritis or broken bones, no severed tendons or cuts or bruises or sores. In short, the exam reveals that you have perfect hands.

But how well can you use those hands? In other words, how strong are they? How much endurance do they have? How coordinated are they, how fast, how dextrous? Can you play the violin like Jascha Heifetz? Can you rattle the piano keys like Jerry Lee Lewis? While a missing thumb would certainly hamper the use of your hands, that your hands are physically perfect guarantees little about your hand fitness.

In this regard, eyes are much the same. During a routine eye exam, the doctor uses lights, mirrors, magnifiers, and an ever-expanding assortment of high-tech instrumentation to measure the power of your eyes and check their every nook and cranny for the slightest trace of disease or abnormality. Even so, when the exam is over and the doctor has pronounced your eyes to be perfect, or has scribbled out a prescription for glasses that will allow you to read the expected lines on the eye chart, we still know nothing about the endurance, speed, accuracy, coordination or flexibility of your seeing. Nothing about your visual fitness.

Visual fitness, then, is more than perfect eyes or perfect glasses. It's harnessing the brain's untapped potential to use those eyes or glasses. As baseball great Yogi Berra once observed, "Ninety percent of hitting is mental. The other half is physical." The same math applies to visual fitness. Having perfect eyes or glasses is half of seeing; using those eyes is the other 90 percent.

Providing that our eyes or glasses are good, visual fitness becomes the capacity to keep up with the information boom on the Internet, to read *War and Peace* without fatigue, to win at tennis or golf, or to navigate a winding road in the dark of night without fear of crashing through a guardrail. Visual fitness is the ability to return home after a day in front of a computer screen and still have the energy left to enjoy your loved ones.

No matter how perfect your eyes or glasses, visual fitness affects you in ways you may not have imagined. The following stories—taken from my twenty-five years of training individuals to use their eyes—share some of the more common ways in which visual fitness affects our lives. Some of the stories involve individual patients, whose names and personal information have been changed to protect their privacy. A few of the visual athletes in the stories are composites built from hundreds of patients whose visual fitness problems, and resultant problems in life, were the same.

Internet Eyes

In the early 1980s, concerns began to arise about what working long hours in front of video display terminals might do to our eyes. Indeed, since that time, surveys have suggested that half of all computer operators complain of eyestrain. With the increasing popularity of the Internet, however, computers are no longer affecting the eyes of just office workers. Consider Bill's story.

Bill is a commercial airline pilot who came to me complaining of blurred vision, especially since he'd begun to spend more time surfing the Net. When he raised his eyes from the computer screen, he needed a few seconds to clear things up far away. More alarming to Bill, the general level of his seeing seemed to be getting worse. He was one of the growing number of computer users whose eyesight is deteriorating as they increase that time spent twenty-some inches away from the computer screen.

In Bill's case, he had become aware of the early signs of the problem before his eye doctor could detect anything wrong. The doctor said that Bill's eyes were perfect and assured him he could see letters on the eye chart he had no business seeing. When Bill continued to insist that there was a problem, the doctor hadn't actually accused him of being neurotic, but Bill had sure felt that way, especially when the professional had looked at him and repeated, "There's nothing wrong with your eyes!"

When I examined Bill, I found his eyes were indeed perfect. He could see from twenty feet away the same line on the eye chart that most of us have to move up to fifteen feet to see.

So why was Bill complaining of blurred vision? To understand his concern, think back to science class in school. Remember the diagram which showed how the eye is like a camera? Images of the world enter the front of the eye, pass through a lens, and focus upside down on the back of the eye. The routine eye exam answers the question, Is the camera broken?

Fortunately for Bill, I asked two additional questions: How well do you take the pictures? and, How well do you develop the film? In other words, I used a number of tests to check his visual fitness.

As it turned out, Bill's cameras may have shamed those produced by Nikon, but his photography skills were sadly lacking. I knew at once that he had some vigorous workouts ahead but, as I consoled him, at least he wasn't going to have to dress for the gym or sweat like a hog.

"Oh, so you mean I have to work out my eyes?" asked Bill. No. Visual fitness takes place in the brain, not the eyes. For this reason, building visual fitness isn't like lifting weights or jogging. It's more like learning to drive a car. Remember your first few times behind the wheel when you had to think about keeping the hood between the lines? Now when you drive, you think instead about where you're going and where the other cars are. Your driving didn't improve because your arm muscles started bulging. It improved because you taught your brain to coordinate those muscles with your eyes.

The same is true when we work visual fitness. We're not building eye muscles. Those muscles are already a hundred times stronger than they need to be to move the eyes. Instead, you could say we're building "brain muscles."

But which areas of the brain are we improving?

Scientists are still resolving that question. If you compare textbooks of brain function today with textbooks of brain function thirty years ago, you'll find that the brain is apparently evolving faster than at any time in history. I'm not going to speculate here about which parts of the brain affect how we see because if the current trend continues, brain function will be understood so differently in ten years that anything I wrote now would no longer apply. For this reason, I'm going to rely instead on an imaginary set of "brain muscles." We'll call these the *visual fitness muscles* and, to keep it simple, we'll agree to imagine they're located somewhere beneath the roof of the skull and above the soles of the feet. Chapters 3 through 9 will introduce you to each of these imaginary muscles so you can begin your visual fitness workouts.

In Bill's case, his flabby visual fitness muscles, further fatigued by the Internet, made it difficult for him to keep things clear. Even though he could see clearly for the few seconds it took to call out a half dozen tiny letters on the doctor's eye chart, his vision was inconsistent, not stable the way it had been before hooking up to the Internet.

As I suspected, part of Bill's dissatisfaction was caused by how good his former vision had been. When I asked him, Bill admitted that a doctor had once been impressed that Bill could see letters twice as small as average patients. Bill was therefore keenly aware when his faraway vision began to blur ever so slightly for a few moments after he used his computer. He was downright concerned when that blur began taking longer to clear up, until his once superior vision had dipped into the high average range.

Had Bill waited to begin his visual fitness program, he might have seen his vision continue downward through average and below until a real eye problem developed and he needed glasses. As soon as Bill found out about visual fitness, however, he began his workouts. By the time he finished, his visual fitness muscles were greased and pumped up for Internet competition. Once more, his vision was superior at all distances, instantly and effortlessly. Once more, he could call off letters twice as small as most of the rest of us can see.

The Night-Driving Nightmare

Some people hate driving at night. Some don't. Look over the following descriptions of day/night driving preference and see where you fall on this eight-point scale:

1. I prefer to drive at night: there's less traffic.
2. I'm equally comfortable driving day or night.

3. I don't mind driving after dark, but night driving does take a little more work.

4. I'd rather drive during the day than at night.

5. Driving at night makes me nervous.

6. I avoid driving at night when possible.

7. Except for emergencies, I don't drive at night.

8. The thought of driving at night makes my heart pound and my hands sweat.

Unless you fall in the first two categories, you may be interested in Darlene's story.

Darlene was a tall, thin redhead in her late thirties. Despite having new glasses, she did everything she could to avoid driving at night. When traveling on business, she always booked morning flights. She didn't want to be caught in a rental car out on strange roads after sunset. At home in Atlanta, Darlene drove a Volvo while her BMW Z3 roadster sat in the garage, unused. The heavy Volvo was the armor she hoped would shield her when she drove at night.

In time, even the Volvo wasn't enough. Darlene's fear of driving even during the day grew into a regular nightmare. She began experiencing full-blown panic attacks, which forced her off the highway to sit tingling on the roadside, trying to catch her breath. Her world was shrinking rapidly.

To understand some of what Darlene was going through, imagine sitting in the darkness of a movie theater, watching a thriller. A man with a knife is stalking an innocent young girl. We know he's out there somewhere, but we can't see him; the director has shrunk the field of view of the camera until all we can see is the girl and the objects touched by her pale fingers: a single chair arm, one side of a door frame. As the menace in the background music slowly builds, we have no idea where the stalker is, no idea when his blade will flash onto the screen.

Well, for Darlene it wasn't the fear of a stalker with a knife that started her hyperventilating. Instead, it was the fear of eighteen-wheelers flashing out of nowhere. But no film director was constricting her field of view. The shrinking of her world was related not to any eye disease, but to her lagging visual fitness.

Darlene isn't alone in her fear of night driving. Although few drivers experience outright panic, many feel a vague nervousness, or perhaps apprehension, when piloting expressways after dark. There are a number, though, who whenever possible scurry home before sunset as surely as if they were trying to avoid Dracula and friends. Many of these day travelers wear glasses. Many don't. But glasses or not, most need to tone up their visual fitness muscles.

During the day, the light entering our eyes provides all the details our brains need to interpret the world before us—assuming we have perfect eyes or perfect glasses. In poor lighting, however, those needed details are obscured. It takes more work to see the signs and lines and edges of the road. This is where visual fitness becomes important.

Good visual fitness allows us to answer two questions: *What is it?* and *Where is it? What-is-it* questions might include, "What does that street sign say?" or, "Is that a Ford or a Buick?" *Where-is-it* questions allow us to answer "Where exactly is that fender?" or, "How fast is that car coming?" While identifying street signs and the makes of cars can be important, judging position and distance could save our lives.

There are some actual eye problems that affect depth perception. Persons who are cross-eyed or have lost vision in one eye are at a disadvantage when it comes to judging depth. There is also a sizable group of people who, despite having perfect eyes or perfect glasses, still have trouble judging depth in the real world, especially at night.

Weakness in this area of visual fitness is the reason why many people are lost without their glasses. The person who can't step out of bed without putting on a pair of glasses does so not because the floor is too small to see, but because of trouble judging distance. Such persons rely on clear vision to see details, and when the details are lost so are they. This same problem with judging distance is aggravated when the light is poor. The less light there is, the fewer cues we have and the better our visual fitness has to be.

In Darlene's case, as soon as she had completed her visual fitness workouts, she was back on the road, her panic attacks a thing of the past. When trucks whizzed by she was no longer concerned, because now she could tell exactly how much space there was between the trucks and herself. Her improvement was so dramatic that she soon had the Z3 back out of the garage. Her night-driving nightmares were over. She no longer needed the armor of her Volvo, even after dark.

Counting the Pages

How long can you read without fatigue or discomfort or loss of comprehension?

While many people can read for hours, not everyone is so blessed. There are a number who start off well, but all too quickly run into problems. For some, these problems consist of sore eyes or headaches. Others see the print begin to dance. Others lose their place. For many, the print never blurs or wavers, but they start feeling as if they'd rather be somewhere else, and then they either have to put the book down or begin rereading sentences to understand what they're reading. Some people read for a short time, then fall asleep. Others have to whisper the words to themselves or say the words out loud so they can take the information in through their ears; for some reason, the information just won't come in through their eyes.

Sarah was one of these people who struggled with reading. She was bright, but even so, she always counted the pages before attempting to read a book. If the book was too thick, she'd return it carefully to its shelf. She similarly avoided paperbacks with small, crowded print.

Sarah had graduated from college with honors and become a teacher. Despite her school record, however, she panicked when asked to read aloud something she had not previously read. She did poorly on timed tests. To maintain her grades, she had worked harder than her classmates and spent more time studying. She had never discovered the joy of reading. For years, in fact, she had secretly wondered if she was slow. She wished that she could sit down and read something for pleasure without it taking forever. She wished that rather than stumbling, she could read out loud with expression.

Sarah's problem was not with reading itself: she could easily sound out words and she knew what they meant. Her problem was fatigue. The longer she read, the more her eyes jerked across the page and the worse her comprehension became. No matter how hard she practiced, her eyes continued to skip over words and even lines, or she'd end up rereading the same sentence twice. She experienced such difficulty that before reading to her seventh grade class, she would take the passages home and practice them so as not to embarrass herself in front of her students.

Because Sarah went for yearly routine eye exams and had worn glasses for much of her life, she had never considered that her eyes might be the cause of her reading fatigue. And, as it turned out, they weren't. Her eyes were fine; her glasses were fine. The problem was with her visual fitness. She put in so much extra effort coordinating her eyes when reading that she fatigued quickly and lost comprehension. Sometimes the effort even caused the print to run together. Each time her eyes moved from one word to another, she would have to struggle to readjust. The readjustment process took so much time and effort that even when she didn't lose her place, she still was forced to pause before reading each word. The required effort was the reason she read "one . . . word . . . at . . . a . . . time."

After several months of work on the use of her eyes, her ease of reading improved dramatically. She now had the confidence to pick up a book and read it orally. Her eyes no longer jerked across the page, and she didn't get tired. She had put off getting her master's degree due to the excessive effort it would involve, but now she had the confidence that she could read fast enough. Best of all, she no longer had to count the pages before reading a book. She finished a Tom Clancy novel of more than 780 pages—she was so excited afterward that she picked up another one and finished it, also without a problem. The last time I spoke to Sarah, she had completed her master's—without having to count the pages.

Too Tired to Get a Life

Have you ever stopped to wonder why you can be so physically drained after a day at work? Why when you get home you're short-tempered or exhausted? Let's look at this. During an average workday, how many stairs do you climb? How many cement blocks do you carry? How many miles do you run? How many push-ups do you push or pull-ups do you pull? How much physical labor do you actually perform?

If your day is spent sitting at a desk, then why are you so exhausted? Which muscles have you overworked? Perhaps your fatigue and frustration have the same source as Todd's.

Todd sometimes felt he was too tired to get a life.

He was a computer programmer. He worked nine to five, but felt that his job consumed his life, at least during weekdays. After eight hours at work, he was too worn out to do much of anything else.

While in front of the computer, Todd experienced headaches and eyestrain, symptoms he relieved with aspirin. Because he wore glasses that were carefully updated each year, Todd assumed his discomfort and fatigue were due to the "stress" of the job. While the aspirin dulled the pain, it did little to maintain Todd's energy level.

By the time he got home, he felt irritable. Had he not lived alone, he would have snapped at anyone whose image further bruised his sore eyes. Instead, he just crashed on the couch for a nap, exhausted.

On Mondays, still refreshed from the weekend, Todd might cut his naps short enough to play basketball with his friends. By midweek, however, the last of his energy stores were drained. His naps would drag on until he forced himself awake long enough to cook and eat before stumbling off to bed. Dating was out of the question. He was just too tired and crabby.

Finally, a friend heard Todd complaining about his eyestrain and general fatigue and suggested he come see me. At first Todd protested that his eyeglasses were new; they couldn't possibly need to be changed yet. When his friend explained, however, that she had suffered from a similar problem despite having the "best" glasses, Todd made an appointment.

When I first met Todd, he was skeptical that I could do anything for him. He insisted that his glasses were new and that they were "correct."

He was right. I found not the slightest change when measuring his eyes for glasses. By the time I had performed a different battery of tests to check Todd's visual fitness, however, he was complaining of the same headache and eyestrain he experienced at work. That I could "provoke" the symptoms suggested that his headaches were related to his visual fitness rather than his sinuses, his blood sugar, or his boss.

When I explained to Todd how he was visually unfit, he was, at first, reluctant to start working out. He claimed he was "too busy." It was simpler just to take the aspirin, which more or less relieved his headaches. At this point, I asked Todd how much energy he had when he got home from work. I explained how, in my experience, people with his level of visual fitness were generally worn out after a day in front of the computer. Either their tempers flared or they just wanted to sleep.

My use of the word *sleep* brought Todd wide awake. He admitted that after a day of using his eyes, he was good for little else but sleeping. He half feared he was sleeping away his life.

He began his fitness program the same week. After his first workout, he left with a headache. That evening, he was so worn out he skipped dinner and went to bed.

Within a few weeks he was completing his exercises with only a mild headache. At work his headaches were also reducing. He was not nearly so tired after work.

Within three months, both his headaches and his fatigue were becoming things of the past. After work, he had the energy to play basketball with his friends. When he did stay home, he once more enjoyed reading, a pastime he'd all but abandoned when visually unfit. He also started dating; the evening hours were no longer past his bedtime.

In short, as Todd's visual fitness grew, so did his interest in everything. He found himself doing better with others. He now had the energy to accept their quirks and enjoy their laughter. He now had the energy to get a life.

Sports Vision

"Keep your eyes on the ball!"

Easier said than done, especially if we're talking about a fastball hurtling in your direction at one hundred miles per hour.

Unless you're a major-leaguer, however, you'll probably not need to develop the use of your eyes to such a level. But what if you'd like to spruce up your golf or tennis game and spend more time winning? What kind of visual fitness will that require?

As I mentioned above, visual fitness helps us answer the questions *Where is it?* and *What is it?* Just as important for sports, though, visual fitness also allows you to answer the question *Where am I?*

Yes. Where are you? Where are your feet? Where are your hands? Where is your body? Good questions to be able to answer, whether we're talking about sports, dancing, or walking around a crowded room without stepping on toes or spilling drinks. Jack and Mary's stories are good examples.

Jack and his wife, Mary, came to me in the hopes of improving their games. Jack was a golfer; Mary played tennis. Both felt that something other than form and technique was holding them back from achieving their potentials.

On drives, Jack was always the last to pick up the flight of his ball and determine if it was long or short. Only his partner's help prevented time-consuming searches. Jack's short game was inconsistent and, other than on his home course where sheer familiarity had bred limited success, his ability to read a green was not the stuff from which legends are made.

Mary had her own problems. On the tennis court, she would take an instant too long to predict the path of the ball. As a result she would arrive too late to use the swing that she had developed during lessons.

Worse yet, playing the net scared her. At that distance, the ball moved too fast to be judged. She feared being struck in the face.

I first met Jack and Mary when they were bringing in their son, Peter, to see me for difficulty with learning to read. During one of the progress exams, Jack shared that not only was Peter's reading improving, so was his ability to catch and hit a ball in Little League. As the topic turned to sports, I could sense that both parents were becoming interested in their own visual fitness.

Signs of improper use of the eyes in tennis include difficulty returning lobbed shots, inconsistent performance despite adequate technique, hesitation in accurately moving the body to the necessary position on the court, and failure to consistently contact the "sweet spot" of the racket. In golf, misjudging the exact distance of the hole (especially during the short game), incorrectly reading the relative elevations and contours of the green, and any tendency to "topping" or hitting the ball "fat" (low) all signal lagging visual fitness.

In both golf and tennis, reduced performance as the game progresses can also be a warning that visual fitness needs work: the extra seeing effort expended can cause premature fatigue and loss of both concentration and accuracy.

When examining Jack and Mary, I found that both had apparently inherited their son's difficulties with visual fitness.

When Jack was in school, he had made satisfactory progress, more because of his intelligence than his ability to use his eyes. He had nonetheless been happy to escape into the reduced visual challenges of a sales position after an exhausting and frustrating four years in college. Despite escaping the books, however, Jack had not escaped his unsuspected visual fitness problem. This problem was now creating stress on the golf course where, while networking, he maintained and secured some of his best accounts. His goal, therefore, was not so much to beat his clients as to avoid embarrassing himself by losing a ball or holding up the game.

Unlike Jack, Mary had been an excellent student in school. She still avidly read novels for fun; indeed, she had that ability to shut out the rest of the world and read no matter how bustling or noisy her surroundings. This shutting out the world had its cost, however. She'd sacrificed her distance vision in favor of improved near performance. The only sign of the intense effort she'd learned to expend on reading was the steadily increasing power of her glasses. Although Mary now wore contacts, the same undiagnosed problem that had silently dogged each step of her way through school was still disrupting her performance—both on and off the tennis court.

When I saw the results of Mary's visual fitness evaluation, it became apparent that she didn't shut out the world only when she was reading. One of my first questions was, "How often do you bump into things?"

The question was so spot on that she laughed. As it turned out, Mary's way of navigating around the kitchen was to walk until she bumped into table, counter, or sink, then to stop. She seldom used her eyes to guide her. Indeed, she admitted that ever since she could remember, she'd had bruises on her hips from running into things.

Perhaps the main difference between seeing to read—at which Mary was excellent—and seeing to hit a ball lies in the realm of "movement." When reading a book, neither book nor head should be moving. By contrast, in tennis, both head and ball are in rapid motion. Visual fitness for sports therefore incorporates the ability to deal with moving targets while our bodies are also moving or unbalanced.

Jack and Mary both did well with their visual fitness workouts. By the time Jack was finished, he believed that he had developed a "natural sense of confidence" because he was "seeing things as they really were." He had golfed with the same partner for a number of years and during that time his partner had always been the one to announce if Jack's shots were long or short. Now, for the first time, Jack, with his speed and depth perception improved, could consistently call the distance before his partner. For the first time, Jack could also expand his view enough to see the elevations and contours necessary to read the green. He was no longer dependent on familiarity with the course to plan his putts. As a result, he felt that his putting had improved tremendously. While he still had no urge to trounce his clients, he was confident that he could offer them a good game without fear of embarrassing himself.

Similarly, Mary's confidence in her tennis game soared. As her reaction time speeded up, she was no longer afraid of getting hit in the face with the ball. Instead she was playing a more aggressive net game. Also, because of her improved peripheral vision, she was aware of the position of the other players without taking her eyes off the ball. Being more confident, alert, and relaxed, she found her attention was freed from "seeing" enough to allow her to place more power into her swing.

While Jack hasn't yet sunk a tee shot, and Mary has yet to be invited to Wimbledon, both are finding greater joy and satisfaction in their games. And now that Mary uses her eyes rather than her hips to navigate around the kitchen, her bruises are a thing of the past.

Seeing to Sleep

Before we continue, I'd like to make a disclaimer: Exercising your visual fitness muscles is not a cure-all, not a panacea.

It only helps when your eyes are open.

Or so I once believed. But consider Steve's story:

Steve came to our office convinced that I could help him. In a magazine, he'd read about a specialized area of visual fitness known as *visualization*, the ability to manufacture and manipulate images in the mind. Some of the more popular uses of visualization are mentally practicing your golf swing or picturing your goals in life. But what if when you close your eyes all you see is blackness? What chance is there of practicing your golf swing in the comfort of your easy chair? And possibly more important, how real can your goals be if you can never see them? Imagine visualizing world peace when all you can see is darkness. What kind of peace would that bring to you or anyone else?

Simply stated, Steve had trouble visualizing. Visualizers can relax and see the pictures in their minds. Nonvisualizers rely instead on a constant stream of inner chatter. It's as if they have to create their thoughts verbally at all times to keep those thoughts from disappearing.

Steve's inability to visualize was causing him problems at work. With no pictures in his mind to support what he was saying, he had difficulty during his sales presentations. Although he typically muddled through, he was afraid of becoming lost or confused, and the more afraid he became, the less he could picture in his mind.

But by the time Steve had finished toning up his visual fitness muscles, he could visualize better than any of the trainers in our office. He was now perfectly at ease with his selling.

During one of his final visits to our office, Steve mentioned that he had always had trouble falling asleep at night. It would frequently take him several hours to doze off. As I just mentioned, those who fail to visualize tend to chatter to themselves. This inner chatter can keep them awake long after their heads hit the pillow. In Steve's case, I suggested that instead of chattering to himself, he begin by picturing the number *100,* and then counting backward, trying to picture the numbers, whenever possible, without verbalizing them.

The next visit, when I asked Steve about his sleep, he was excited to tell me that it had taken him only ten minutes before his mind had begun to wander toward a deep and peaceful slumber. It made sense: how can you count sheep if you can't see them? Steve's visual fitness had improved so much that he could now see his way not only to sell, but to sleep.

Wear Them—You'll Get Used to Them

Do you wear glasses or contact lenses? If so, answer the following questions: Do your glasses help you see but give you a headache or cause your eyes to pull? How dependent are you on your glasses? Do you wear them only when you need them, or do you need them all the time?

While the power of your glasses is largely based on the structure of your eyes, how well you get along with those glasses relates to your visual fitness. True, if your eyeglasses are as thick as heavy ashtrays, you're going to be a bit dependent on them. But what if your glasses aren't all that strong?

When it comes to weaker glasses, there is quite a variation of eyeglass dependency. I might, for example, examine four people with the same relatively weak eyeglass prescription and find that one person puts the glasses on before getting out of bed in the morning, one wears the glasses only for driving, one wears them not at all, and one wears them all the time but nonetheless complains of eyestrain. Sure, if your glasses are as thick as those water bottles used in the cooler at work, you may never be able to throw them away. But even with those thick glasses—providing your prescription is correct—how well you function will depend primarily on your visual fitness.

If you're one of the many who can't live with your glasses and can't live without them, consider Jane's story.

Jane was in her thirties when she began to experience some blurred vision when reading. She went to the eye doctor, who found she needed a relatively weak pair of glasses.

When Jane got the glasses home, however, she was discouraged to find that although the glasses made the print clearer and blacker, they seemed to distort the world and strain her eyes. When she called the doctor to report her symptoms, he assured her that her situation was nothing out of the ordinary. To get used to the glasses, she would merely have to continue wearing them. Within a few weeks she'd be "doing just fine."

Such advice is generally true. Although new glasses may make the walls appear bowed and the floor to slant, these warped perceptions normally pass.

Jane, however, wasn't "normal." She struggled with her glasses for a month before returning to the doctor for a reexamination. He found a minor change and ordered a new set of lenses. Jane found the new set no better than the first. By the time of her next visit, the doctor was beginning to share her frustration. But it wasn't until her fourth visit that he gave up. Exasperated, he demonstrated for Jane how she could read the tiniest letters on his chart. "This is as close as we can get," he finally announced. "I'm afraid there's nothing else I can do for you." He added, "Are you sure you're not under any other stress?"

While Jane silently allowed that the doctor himself was apparently becoming quite stressed by her problem, she could recall no particular stress in her own life outside of being unable to comfortably read with or without her new glasses.

Not willing to give up, she tried another doctor, but with much the same result. Finally she heard about visual fitness from a friend. Four months later she had toned up her visual fitness muscles and was reading comfortably, with or without her glasses.

The point here isn't that better visual fitness will allow you to throw away your glasses (sometimes it will and sometimes it won't). The point is that if your eyes are healthy and you're struggling even with your new glasses, then visual fitness may be your only key to comfortable, efficient seeing.

Developing Your Visual Fitness

So how about you? Whether you have perfect eyes or perfect glasses, is there some area of your life which could be improved by seeing clearer, bigger, deeper, stronger, quicker, or smarter? How about enhancing your reading, driving, golf, or tennis? How about reducing eyestrain, boosting your energy level, picturing your goals, or falling asleep at night? These are but a handful of the many ways in which better visual fitness can enhance your life. But even if your life is already as perfect as your eyes or glasses, visual fitness could still be of use.

Try the following experiment. Close your eyes and look around the room. What do you see?

Now open them.

See better? Things a little brighter?

Well, what if it were possible to open your eyes a second time? I don't mean close them again and reopen them. I mean, what if it were possible to open them wider, somehow, and see the world brighten again as much as it did the first time? What if it were possible to build upon your 20/20 Volkswagen Bug way of seeing, and open your eyes as wide as the throttle of a gleaming new Porsche or Mercedes-Benz?

Would it be worth seven minutes a day?

Visual fitness—it's not just about eyes. It's about seeing big enough to change our lives and small enough to zero in on the details of reading and computer work. It's about seeing the forest or the pine needles or both as success demands. It's about being able to learn from what we've seen in the past and having the ability to create the future. Visual fitness is about opening not only our eyes but our minds to the possibilities and opportunities of the world.

To discover the many ways in which visual fitness could expand your life—ways which you may never have imagined—turn the page and begin your Visual Fitness Self-Assessment.

2

Your First Step:
The Visual Fitness Self-Assessment

We are now going to assess how a need to improve your visual fitness could be impacting your life. For the purposes of this assessment, we'll assume that you've had a routine eye exam in the not too distant past and you already have perfect eyes or, at least, perfect glasses.

You are going to score yourself on eighty statements that could provide clues about your level of visual fitness. As you rate yourself, use the following five-point scale:

0 The statement *never* applies.

1 The statement *seldom* applies.

2 The statement *occasionally* applies.

3 The statement *frequently* applies.

4 The statement *always* applies.

Read each statement. In the blank next to it, write the number that best describes your experience. If you wear glasses, consider the statements as if you were wearing your glasses. For instance, for the statement, "When you read, the print blurs," consider only those times blurring occurs while you are wearing your reading glasses, not when you have your glasses off. When you complete each section, add up your scores for that section and place the total on the line next to the words *Section Score.*

The Visual Fitness Self-Assessment

READING

1. When you read, the print blurs. ______
2. When you read, the print runs together. ______
3. When you read, the print looks unsteady or dances. ______
4. Reading gives you eyestrain or headaches. ______
5. Reading puts you to sleep. ______
6. You avoid reading for fun. ______
7. You avoid longer books. ______
8. You avoid books with smaller print. ______
9. When you read, you get the feeling that you'd rather be somewhere else. ______
10. You rapidly fatigue and lose comprehension when reading. ______
11. You have to whisper to yourself when reading. ______
12. Reading gives you an upset stomach. ______
13. You lose your place and skip or reread lines. ______
14. You're afraid to read out loud in front of other people. ______
15. Reading takes too much effort. ______
16. You read "one . . . word . . . at . . . a . . . time." ______

17. You have to reread sentences to understand what you are reading. ________

Reading Section Score: ________

DRIVING

18. You get eyestrain or headaches when you drive. ________
19. You get carsick, especially when sitting in the back seat. ________
20. You rapidly fatigue when driving. ________
21. You dislike driving at night. ________
22. You have difficulty judging how far away other cars are. ________
23. You find parallel parking difficult. ________
24. You have to look twice because you can't trust yourself to see things correctly the first time. ________
25. You have difficulty telling how fast other cars are moving. ________
26. You have trouble seeing road signs. ________
27. It makes you nervous to drive when traffic is heavy. ________
28. It makes you nervous to drive on the freeway. ________
29. At night, the taillights ahead of you seem to double up. ________
30. You get lost easily when driving. ________
31. Your worries about driving limit your activities. ________

Driving Section Score: ________

WORK

32. You have more trouble with computer and desk work as the day goes on. ________
33. You have to schedule your computer and desk work in the morning when you're fresh. ________
34. Your productivity goes down as the day progresses. ________

35. You get eyestrain or headaches during computer or desk work. ________
36. Your stomach gets upset during computer or desk work. ________
37. You reverse numbers at work, such as seeing 36 for 63. ________
38. You have to check your work for errors constantly because your eyes play tricks on you. ________
39. Your computer or desk work takes longer than it should. ________
40. You put off your desk work and instead spend your time talking, either face-to-face or on the phone. ________
41. You'd have second thoughts about a promotion if it meant more reading or desk work. ________

Work Section Score: ________

RELATIONSHIPS

42. You have trouble maintaining eye contact when speaking to someone. ________
43. You feel like backing away when a person is speaking to you. ________
44. You feel as if you need to move right up next to people when they are talking to you. ________
45. You are too tired to enjoy your friends or family after a day of using your eyes. ________
46. After a day of using your eyes, you are irritable or short-tempered. ________
47. Sore eyes or headaches interfere with your relationships. ________
48. Desk work drags on forever so you have little time left to enjoy your friends or family. ________
49. The effort it takes you to read has kept you from going back to school and is therefore limiting your income. ________

50. Your worries about driving limits the number of activities in which you or your children get to participate. ________
51. Your reading ability affects your confidence around others. ________
52. Your driving ability affects your confidence around others. ________
53. Your coordination affects your confidence around others. ________

Relationships Section Score: ________

SPORTS

(If a statement applies to a sport you don't play, give yourself a score of 0.)

54. When you exercise, you prefer walking, running, swimming, calisthenics, or lifting weights rather than visual activities such as baseball, tennis, or golf. ________
55. When it comes to ball sports, you're a klutz. ________
56. You've always avoided participating in ball sports. ________
57. It's hard to catch or hit a ball. ________
58. When playing golf, your short game is more difficult. ________
59. When playing golf, it's not easy to read the green. ________
60. When playing golf or tennis, you consistently hit long or short. ________
61. In whatever ball sport you play, it's harder to maintain your concentration the longer the game continues. ________
62. In any ball sport, you're not as good as your technique would predict. ________
63. In tennis you have trouble with returning lobbed shots. ________
64. In tennis, you have more difficulty at the net than at the baseline. ________

Sports Section Score: ________

COORDINATION

65. It bothers you to walk down stairs. ________
66. You bump into things. ________
67. When dancing, you have two left feet. ________
68. It makes you nervous to walk in a crowd. ________
69. You're clumsy. ________
70. You trip and stumble if you're not careful. ________

Coordination Section Score: ________

GENERAL

(If you don't wear glasses, give yourself a score of 0 next to the statements that mention glasses.)

71. Things are blurry for a moment when you look up from reading or computer work. ________
72. You get headaches or eyestrain when you use your eyes for careful seeing. ________
73. Your stomach gets upset after you use your eyes for careful seeing. ________
74. Things blur in and out of focus when you look at them. ________
75. Your glasses give you headaches or eyestrain even though you need them to see. ________
76. Wearing your glasses makes you sick to your stomach. ________
77. Since you started wearing glasses you find yourself avoiding reading. ________
78. It makes you nervous to search the crowded shelves in the grocery store. ________
79. It's hard to fall asleep because you keep chattering to yourself. ________

80. You have difficulty seeing words in your mind to spell them. ________

General Section Score: ________

Scoring Your Assessment

In the spaces provided below, copy your seven section scores. Add them together to obtain your Visual Fitness Score.

Reading Section Score: ________

Driving Section Score: ________

Work Section Score: ________

Relationships Section Score: ________

Sports Section Score: ________

Coordination Section Score: ________

General Section Score: ________

Visual Fitness Score: ________

So how do you think you did? Did you come across statements that you've never considered might be related to the use of your eyes? Perhaps you're even a little skeptical that in your case certain statements have anything to do with the use of your eyes. Well, you may be right. Let's consider headaches, for instance. They appear five times on the assessment and could certainly influence your score.

Headaches

Just as headaches can be caused by eye problems—such as a need for new glasses—headaches can also be caused by a need to improve your visual fitness even when a routine eye exam proves your eyes or glasses are perfect.

Obviously, however, not all headaches are caused by inefficient use of the eyes. Visual headaches, for instance, don't wake us up in the middle of the night. Similarly, if you wake up in the morning with a headache that didn't accompany you to bed the night before, that headache couldn't be due to a problem with using your eyes.

Sinus problems, food allergies, illnesses, low blood sugar, brain tumors, and heaven knows what else can all cause headaches. And certainly, even if you have headaches caused by a need to improve your visual fitness, you could still have headaches caused by other more serious problems as well. For this reason, if you suffer from headaches you should see your physician and rule out any health-threatening physical concerns.

On the other hand, if your doctors have ruled out any physical reason for your headaches, and they are telling you that your headaches are "all in your head" or due to "general stress," then a need to improve your visual fitness could be involved. Indeed, perhaps your stress isn't "general" at all. Perhaps it's "visual."

If you are careful to note when your headaches occur and find them to be worse when you use your eyes for careful seeing and better when you use your eyes for casual seeing, they could well be visual. Computer work, desk work, reading, studying, driving, and watching television or movies are all examples of tasks that can involve careful seeing and trigger visual headaches. Such headaches are typically of the muscle-tension variety. They start in the forehead or around the eyes, or they cause the muscles in the neck or the back of the head to cinch up.

A good way to understand such headaches is to try the following demonstration: Lay your hand palm down on your leg. Snap your ring finger up in the air as forcefully as you can. Notice how when you raised your ring finger, you simultaneously moved other fingers as well. This movement of the other fingers is an example of "overflow." You ordered your ring finger, "Move up in the air." Your brain, however, moved other fingers as well. The same thing happens when you get a visually related muscle-tension headache. You order your eyes, "See that!" Your brain, however, flexes not only your eye muscles, but surrounding muscles as well. The result: a nagging headache caused by the overstimulation of the muscles in your forehead or the back of your head and neck.

Perhaps the best way to determine if your headaches are caused by a need to improve your visual fitness is to perform the exercises listed in chapters 3 and 6. If those exercises cause the same type of headaches that normally bother you, then it's likely that your visual fitness program will reduce your headaches. If, however, the exercises cause no headaches, or they cause a different type of headache than the one you're trying to solve, then it's far less likely that a visual fitness program will reduce your headaches.

We can apply the same reasoning to many of the other statements on the assessment. Sometimes for instance, clumsiness is just that: clumsiness. But clumsiness can't explain why a golfer with good form can't read a green, or a tennis player with good form can't connect the ball with the sweet spot of the racket. Similarly, general fatigue can be due to illness or lack of sleep, but such physically caused symptoms don't miraculously disappear when we get away from the computers, books, or desk work.

In the same way, psychological factors can cause anxiety during driving, walking down stairs, or shopping in crowded supermarkets. At work, there can be bosses and coworkers who, by their constant negativity, cause stress. There are even long hours and unreasonable deadlines which can further exhaust us. But what's important to understand is that the symptoms in the assessment *can* be caused by a need for better visual fitness.

In the chapters that follow, you'll learn more about how some of the assessment statements made it onto the list. In the meantime, let's look at your scores.

Evaluating Your Self-Assessment

First of all, your grade:

0 to 10 Excellent—wow!

11 to 20 Good—still room for improvement.

21 to 40 Fair—but plenty of room for improvement.

Greater than 40 Time to start your workouts—yesterday!

In adults at least, the number-one adaptation to poor visual fitness is to avoid those areas where the weakness exists. Those whose visual fitness affects reading are likely to be found earning a living with their mouths or hands rather than their eyes. Those whose visual fitness affects coordination, driving, and sports, may feel more at ease behind a desk or computer—which, as we will learn later, may further add to the coordination problems.

But what about the salesperson who moves up to sales manager and suddenly has more desk work? Or the visually inefficient reader who loves books, or the visually uncoordinated athlete who wants to play tennis? These are the types of situations in which visual fitness begins to bring on the more traditional symptoms of stress, frustration, headaches, eyestrain, or blurred vision. For this reason, the question to ask may not be, Does reading give me eyestrain? Instead, the question to ask is, Are there areas of my life that I once enjoyed but now I'm beginning to avoid?

Visual Fitness and Quality of Life

There are hundreds of studies that show how eye exercises can improve the use of our eyes. You'll find a sampling of these in Appendix B.

Unfortunately, all too often such studies are conducted from the doctor's point of view. They show how many degrees a cross-eyed patient's eyes straightened, or how before eye exercises a patient could read one size of letters on a chart and after the exercises the patient could read smaller letters. While such studies are necessary, they tell only part of the story.

Over the past decade and a half, doctors have become interested in not only how treating certain parts of our body affect the organs themselves, but how such treatments affect the patient's quality of life.

In 1995, I explored this issue in a study published in the *Journal of Optometric Vision Development*. The study wasn't concerned with how well visual fitness training works; there are enough of those already. Instead I was interested in finding out how visual fitness training affects quality of life—what kinds of changes people were seeing and interested in enough to report. With this in mind, I asked 838 people to answer a single question in writing: What changes have you seen since beginning visual fitness training? (I actually used a different term for "visual fitness training," but we'll get to that in chapter 11 when we learn more about visual fitness trainers themselves.)

The range of responses was little short of amazing. Of the sixty-five changes that at least four different patients mentioned, only fourteen of the changes had to do with eyesight or the alignment of formerly crossed eyes. Forty-nine changes had to do with other areas. One hundred sixty-two patients mentioned "fewer headaches." Almost seven hundred described improvements in reading comprehension, speed, or enjoyment. Almost five hundred mentioned behavioral changes such as improvements in attention, self-confidence, and family relationships. One hundred twenty-four saw improvements in sports, driving, or coordination.

Whether you can hope to enjoy improvements similar to those in the study is another question. As my Arkansas-raised gymnastic coach used to say, "The proof of the pudding is in the eating." If you begin your visual fitness workouts and your symptoms begin to reduce, you'll know you're on the right track.

How to Get Started

If you picked up this book in an eye doctor's office, then that doctor either trains visual fitness or knows where to find someone who does. Perhaps you were even referred to this doctor because of an actual eye-muscle problem such as strabismus (in which one eye drifts in toward the nose or out toward the ear) or convergence insufficiency (in which the eyes cannot work together at closer distances). Eye-muscle problems could dramatically affect the score on your Visual Fitness Self-Assessment. If you have such a condition, you're going to need a doctor's help.

If your routine eye examination is normal, but you're lucky enough to be in the office of a doctor who trains visual fitness, there are still advantages of having your own visual fitness trainer. The trainer will select from hundreds of procedures rather than be limited to the dozen or so outlined in this book. The trainer will tailor your program to fit your exact needs and adapt each procedure to fit your style. In addition, you'll have a coach, someone who stays on you to get you through the workouts and ensure that you're successful.

On the other hand, if you found this book in a bookstore, and you're eager to learn more about visual fitness and get started on improving your own, then you'll find what you need in the pages that follow. The procedures given should help anyone with perfect eyes or perfect glasses to make his or her seeing clearer, bigger, deeper, stronger, quicker, and smarter. If later, after you're already seeing improvements, you want a visual fitness trainer to help you progress even further, you'll be able to learn more about them in chapter 11.

One more thing before we start: If you've completed your own assessment and you have a child whom you now suspect needs help with visual fitness, I suggest you read my book *When Your Child Strug-*

gles: The Myths of 20/20 Vision for more information and then find a visual fitness trainer. The exercises in the following chapters were not designed with the needs of children in mind.

This said, let's turn the page and begin to learn how to flex your visual fitness muscles.

3

VISUAL FITNESS SKILL I: CLEAR-SIGHT

How "Little" Can You See?

About the time of the American Civil War, Dr. Herman Snellen designed a chart for testing eyesight at a distance of twenty feet. To this day, charts modeled after Snellen's still adorn the office walls of doctors and school nurses.

Nineteenth-century visual scientists had measured the parts of the eye and calculated that we should be able to see letters a little less than half an inch tall when twenty feet away. Those who could see these perfect letters were said to have perfect eyesight—20/20 eyesight, to be exact.

Since those pioneering efforts at measuring vision, we've discovered lead-free gasoline and cell phones, we've tamed polio and walked on the moon. But, despite the latest laser surgeries and tinted contact lenses, despite the ever-increasing visual demands of the Internet, the bottom line of routine eye care has changed little from the bottom line of the eye chart that Snellen conceived almost a century and a half ago.

Seeing "Little" Far Away

The best way to understand Dr. Snellen's chart is to take a look at an eye chart of our own. Obviously our sample chart (on page 35) is not the same size as those found in doctors' offices, but it will serve to increase your understanding.

Beside each line of letters, there is a number printed, which tells us from what distance we're expected to read that line. For instance, next to the big *E*—which, as you know, graces the tops of most eye charts—there's a *200*. If you had perfect eyes or perfect glasses, you'd be able to see this *E* when standing two hundred feet away from the doctor's chart. If you had to move up to twenty feet to read the *E*, you would be said to have "20/200 eyesight"; you'd be seeing from twenty feet what you're supposed to see from two hundred. The top number of the fraction refers to the distance at which the test is performed. The bottom number refers to the distance at which those with "perfect eyes" should see the line.

If your eyesight were worse than 20/200, you could still measure it by moving closer to the chart. If you had to move up to ten feet to see the letter marked *200*, your eyesight would be 10/200, which is equivalent to 20/400. If you had to move up to five feet, you'd have 5/200 or 20/800 eyesight. In other words you'd be able to see at twenty feet what Snellen predicted you'd see at eight hundred feet; you'd be about as blind as a bat with a monkey wrench jamming his radar system.

Let's move on down the chart. In the next three lines we find the letters you should be able to see from one hundred feet, eighty feet, and sixty feet. If you needed to move up to twenty feet to see one of these lines, your acuity would be 20/100, 20/80, or 20/60, depending on which line was the smallest you could see. If you could see 20/60, you could drive a car in my home state of Georgia.

Farther down the chart you'll find the lines marked *40* and *30*. If you could see the letters beside these numbers at twenty feet without glasses, you'd have 20/40 or 20/30 uncorrected eyesight, eyes good enough to pass the driver's license requirements in most other states. Interestingly, when you consider that Georgia insurance rates are lower than many states with tougher eyesight requirements, it becomes apparent that there's more to seeing to drive than reading the "proper" letters on an eye chart. As we will learn in later chapters, safe driving relies on many different visual fitness skills.

200 E

100 S H

80 V F

60 E L

50 F R K

40 Z T N

30 H A F B

25 K O V F

20 T S L N P

15 E Z A F L C

10 Q A B T E O Z L

Anyway, back to our chart. Finally, we reach the *20* next to the letters *T, S, L, N, P.* This is the famous 20/20 line—letters one-tenth the size of the 20/200 *E.* If you could see these letters from twenty feet, as Herman Snellen probably did, you'd have "20/20 eyesight."

So now that we've reached perfection, we're done with the eye chart, right? Not really. At this point, you may be asking yourself, If 20/20 is "perfect eyesight," why are there two smaller lines? Is perfect eyesight somehow imperfect? Are there also "more perfect," and "most perfect" eyesight?

In a way, yes. Most younger adults with perfect eyes or glasses can see at twenty feet what those nearsighted nineteenth-century scientists saw at fifteen feet. That is, most younger adults have 20/15 eyesight. Indeed, Bill, the pilot we met in chapter 1, walked in complaining even though his eyesight was 20/15. No wonder the doctors who focused on Bill's eyes rather than his visual fitness suspected he was a bit odd. As it turned out, Bill wasn't happy until he could read the very bottom line of the chart. That line is composed of letters twice as small as those in the 20 line. If you had 20/20 vision, you'd need to move up to ten feet to see the *Q, A, B, T, E, O, Z, L* on that line. If you could see those letters from twenty feet, then you'd have 20/10 vision and be well on your way to Clear-Sight.

Seeing "Little" Up Close

We use a different chart to measure eyesight up close. Take a look at the Near Testing Eye Chart on page 37. This chart is true to size and, unlike the chart used for testing distance eyesight, the numbers beside each line on the near chart do not refer to any particular distance. Instead, when we hold the chart at sixteen inches, the numbers refer to "equivalent" eyesight. In other words it should be as hard to see the 20 line on the small chart at sixteen inches as it is to see the 20 line on the big chart at twenty feet—provided, that is, you have perfect eyes or perfect glasses. If you can read the bottom letters on the near acuity chart, you're even closer to reaching Clear-Sight—but not quite there.

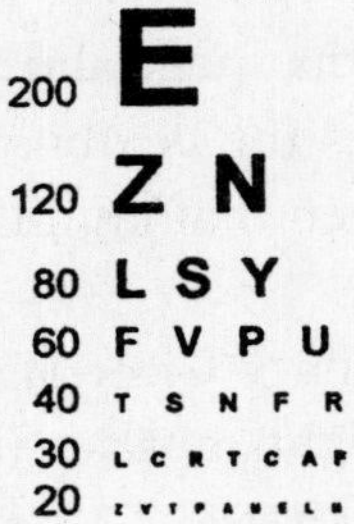

200 E

120 Z N

80 L S Y

60 F V P U

40 T S N F R

30 L C R T C A F

20 Z V T P A N E L N

Near Testing Eye Chart

Clear-Sight

The skill of Clear-Sight requires more than seeing far or near. Even if you can see the tiniest letters on the eye charts at twenty feet and sixteen inches, this is no guarantee of Clear-Sight. That you can see to read a half dozen letters says nothing about your ability to read an entire page or your ability to read that page and then focus on an object in the distance. Clear-Sight requires the flexibility to see clearly when changing viewing distances or when maintaining focus at the same distance for longer periods of time.

In addition to our imaginary Clear-Sight muscle in the brain, there are actual, physical Clear-Sight muscles (called ciliary muscles) inside your eyes—one muscle in each eye. Those muscles relax when we look far away and contract when we look up close. If our eyes are like cameras, then our Clear-Sight muscles are the key to focusing the camera lenses. It's not enough, however, to be able to keep the camera focused clearly for the few seconds that it requires to capture a line of six or eight letters on an acuity chart. Clear-Sight requires four different skills: 1) the *accuracy* to see tiny letters, 2) the *power* to see up close, 3) the *flexibility* to change focus from far to near or near to far, and 4) the *endurance* to see clearly for longer periods of time.

Let's look at each of these four skills separately.

ACCURACY

The size of the letters we can see on the eye chart gives us a good indication of the *accuracy* of our Clear-Sight muscles. Surprisingly, we do not need to focus our Clear-Sight muscles perfectly to see the 20/20 letters, because 20/20 eyesight—despite its reputation—is far from perfect. Letters as large as the 20/20 letters gives our Clear-Sight muscles some leeway—even if we tighten the muscles in our eyes too much, or not enough, we can still make out the letters during an eye test. It is common to find both children and adults who complain of "blurred vision" despite measuring 20/20. On the other hand, seeing the 20/10 letters (letters that are only half the size of the 20/20 letters) requires complete accuracy of our Clear-Sight muscles. If those muscles fail to contract exactly the right amount, we cannot see letters as small as the 20/10 letters.

POWER

We determine the *power* of our Clear-Sight muscles by measuring how close we can see tiny letters. Most fifteen-year-olds with the normal use of their eyes have the power to see tiny letters as close as three or four inches. From the teens on, however, that power begins to wane. Due to the normal aging process, sometime in our thirties we lose half of our Clear-Sight power and our near focus goes from four inches to eight inches. Over the next decade, we lose half our power again, and our focus creeps out to sixteen inches. The next time we lose half our Clear-Sight power—usually between the ages of forty and forty-five—our focus jumps out to thirty-two inches and suddenly our arms are too short to allow us to read. Although a number of the patients I've seen have claimed that the problem was with their arms rather than their eyes, all have preferred reading glasses to the prospect of having their arms surgically lengthened until their knuckles dragged on the ground.

Those with mature eyes, however, are not the only ones whose Clear-Sight muscles need a boost of power. Occasionally I see young adults or even children who lack the Clear-Sight power they need to read or study without fatigue. To get an immediate idea of your own Clear-Sight power, cover an eye with one hand and move the Near Testing Eye Chart on page 37 as close as you can and still make out the letters in the 40 line. Change eyes and repeat the process. If you wear reading glasses, wear them for this self-test. After you've done your Clear-Sight exercises for two weeks, recheck your Clear-Sight power in the same way and see if there has been a change.

FLEXIBILITY

Those who see things blur when they first look up from their books or computers are experiencing difficulty with the *flexibility* of their Clear-

Sight muscles. If your hobbies or job requires you to shift focus from one distance to another, and this shifting isn't almost instantaneous, then you're probably throwing away hours waiting for things to clear. For example, the student who takes longer than others to shift focus between the chalkboard at the front of the room and the notebook on her desk will take longer to complete work that's written on the board.

ENDURANCE

As I mentioned above, it's not enough to be able to call out half a dozen letters on an eye chart. You have to be able to maintain that clarity over time even if you can read 20/10. Those who have trouble with Clear-Sight *endurance* have trouble sustaining vision while reading or driving. Commonly, they experience blurred vision, sore eyes, and even headaches when using their eyes for extended, careful seeing. The ease with which you can perform the Clear-Sight exercises described in the second half of this chapter will give you a good idea about both your Clear-Sight flexibility and endurance.

In the first chapter, the story about Bill's "Internet eyes" was an example of how 20/20 eyesight is a good starting place, but only a starting place. Having 20/20 eyesight doesn't guarantee that we can keep the print effortlessly clear when reading. Having 20/20 eyesight doesn't guarantee that things are clear in the distance when we first look up from a book or computer monitor, or that we don't suffer from headaches or upset stomachs when we read. Having 20/20 eyesight doesn't guarantee that we can read street signs at night or see as well as those around us.

At the very best, all that 20/20 eyesight guarantees is that we can stand twenty feet away from a nineteenth-century eye chart and keep things clear long enough to read six or maybe eight letters.

"So," you may ask yourself, "why settle for 20/20 eyesight?"

My answer is, of course, "Yes, why?"

Why settle for sore eyes or headaches during computer work? Why settle for the extra effort that silently fatigues us when we read and leaves us worn out by day's end? Why settle for straining to see street signs during night driving? Shouldn't that Civil War eye chart have been buried in Arlington Cemetery well before the close of the twentieth century? In short, why settle for cotton-gin seeing during the age of the Internet?

Well, if you're ready to bring your visual fitness into the twenty-first century, then it's time to begin by learning to flex your Clear-Sight muscles.

Before starting, however, let me caution you: Like any other exercise, putting these muscles through their paces may initially cause some aches and pains. Your eyes may tug and burn. You may feel vaguely headachy. Even your stomach may protest because the same involuntary system of nerves that controls your focusing also controls your stomach. If, however, you hang in there and continue putting in your seven minutes per day (three and a half minutes per eye), the aches and pains you experience will gradually disappear, not only when you do your exercises, but all day long, whenever you use your eyes.

Accuracy. Power. Flexibility. Endurance. These are, after all, what visual fitness is about.

Enough said. Let's begin. Even if you're reading through the entire book first and will come back and develop your visual fitness program later, give Clear-Sight I a quick try to learn a little more about how your Clear-Sight muscles work. Or, if you'd rather not have to get up from your seat, you can try Clear-Sight III—just go easy.

When reading these exercises, or any in the following six chapters, don't read the exercise all at once. Instead, perform each step of the exercise before reading the next step. By actually doing the exercise, rather than just reading about it, you'll avoid confusion.

Your Clear-Sight Workout

Clear-Sight I

There are three Clear-Sight Charts. The chart with the large letters is the Clear-Sight Far Chart (see page 45). The two with the smaller letters are Clear-Sight Near Charts A and B (see pages 47 and 48). Cut them out or photocopy them.

If you don't need glasses, great! You won't need any for this exercise. If your doctor has prescribed glasses for full-time wear, use them. If your glasses are weak, and your doctor has said you can wear them when you want to, and you'd rather not wear them, then try this exercise without them. If your prefer to wear them, then wear them.

Do the exercise as follows:

1. Attach the Clear-Sight Far Chart to a wall that is well lighted.

2. Back as far away from that chart as you can while still being able to read it easily—probably around six to ten feet.

3. Hold the Clear-Sight Near Chart A in your right hand.

4. Cover your left eye with your left hand. Don't press on your eye. Cup your palm so that you can leave both eyes open.

5. Move Chart A, the near chart, as close as you can read it comfortably—probably between six and ten inches. If you are older than forty you may have to begin at sixteen inches.

6. In this position (with your left eye covered, the far chart as far away as you can *comfortably* read it and the near chart as close as you can *comfortably* read it), read the first three letters on the far chart: *E, F, T.*

7. Move your eyes to the near chart and read the next three letters: *Z, A, C.*

8. Continue reading all the way down the charts, reading three letters far then three letters near. Far chart: *V, D, P.* Near chart: *A, B, Y.* Far chart: *O, K, E.* Near chart: *W, F, M.*

9. When you have completed the charts with your right eye (or have spent three and one-half minutes, whichever comes first) change the near chart to your left hand. Cup your right hand over your right eye. Again, leave the covered eye open and don't press on it.

10. Read down the charts with your left eye, three letters at a time, the same way you did with your right eye: *E, F, T,* far: *Z, A, C,* near; and so forth.

As you work Clear-Sight I, you may notice that when you first change between the charts it takes a moment for you to focus on them. Each time you look far away, you are relaxing your Clear-Sight muscles. Each time you look up close you are contracting them. The more quickly you can refocus the chart, the better your *flexibility.* The longer you can continue the exercise comfortably, the greater your *endurance.*

The reason you are working with the charts positioned at a comfortable distance is to get used to contracting and relaxing these muscles without straining your eyes. At least at the start, work no more than seven minutes a day on this exercise—three and a half minutes for each eye. Gradually, step back a little farther from the far chart. Gradually hold the near chart a little closer. Once you can perform this exercise without discomfort, you are ready to move on to Clear-Sight II.

Clear-Sight II

On Clear-Sight I, your goal was to learn to shift your focus comfortably and quickly from one distance to another, an ability that will also help you to sustain focus when reading or driving, or whenever seeing details is needed. On Clear-Sight II, you want to extend your range of clear seeing further, increasing your *power* and *accuracy*.

On Clear-Sight II, follow the same ten steps you did for Clear-Sight I, with the following exceptions: In step 2, move farther back from the far chart until you can barely make out the letters. For instance, if during Clear-Sight I you could easily see the letters while standing ten feet away from the far chart, now move back to twelve feet. Keep moving back as you improve until you can stand at twenty feet. Similarly, in step 5, instead of holding the near chart where you can comfortably read it, now hold it several inches closer, where you have to put in effort to make out the letters. Work until you can read the near chart four inches away from your eyes. If you are over the age of forty, it may be impossible to move the near chart to four inches. You may have to work at six inches or ten inches, or even sixteen inches where your bifocal power allows. You'll have to discover the correct distance for yourself. Just be sure you are holding the near chart so close that you can just barely make out the letters. As long as you push yourself, you'll expand your range of clear seeing compared to how you now see.

When you can stand ten feet away from the far chart and clearly read its letters, you'll have about 20/20 eyesight. If you can back up to a little more than thirteen feet and still see the letters, your eyesight is approximately 20/15. And finally, if you can read the far chart from twenty feet away, you've doubled your eyesight compared to those shortsighted scientists of the nineteenth century; that is, your eyesight is 20/10; you can see at twenty feet what they had to move up to ten feet to see.

Clear-Sight III

Clear-Sight III is designed to further enhance your accuracy, power, flexibility, and endurance within arm's reach. The procedure could easily be done at your desk at work.

Use the Clear-Sight Near Chart B. If you have reading glasses, be sure to use them for this exercise. If Chart B is too small to see even while wearing your reading glasses, use Chart A instead.

Perform the following steps.

1. Cover one eye.

2. Move Chart B as close as you can while still being able to comfortably make out the letters.

3. Gently blink your eyes and keep breathing while you see if you can move the chart any closer while maintaining focus.

4. Next hold the chart out as far away from your eyes as you can while still easily reading it—to arm's length if possible.

E F T Z A C V D P

A B Y O K E W F M

N F O P D V C T H

T B X A O C Z N L

V C X C W N S O P

A Y T F Z Q D P C

L B S W F K X T N

Z N A O C Q P F Y

Clear-Sight Far Chart

5. Gently blink your eyes and keep breathing while you see if you can move the chart any farther away while maintaining focus.

6. Continue moving the chart between the near and far positions for three minutes. Be aware of how long it takes you to refocus the chart.

7. When you're finished with one eye, cover it and repeat the above steps with your other eye for another three minutes.

8. Finally, with both eyes open move the chart between the close and far positions for the last minute.

Once you've completed Clear-Sight I, you can alternate Clear-Sight II and III by spending seven minutes on Clear-Sight II the first day and seven minutes on Clear-Sight III the next.

Your Exercise Schedule

I'll say more in chapter 10 about your exercise schedule, but if you'd like to get started now with your Clear-Sight exercises, set aside seven minutes a day, the same time each day, and have a go at it. If you do, you'll be on your way to better visual fitness even before you finish reading this book.

E F T Z A C V D P

A B Y O K E W F M

N F O P D V C T H

T B X A O C Z N L

V C X C W N S O P

A Y T F Z Q D P C

L B S W F K X T N

Z N A O C Q P F Y

Clear-Sight Near Chart A

E F T Z A C V D P

A B Y O K E W F M

N F O P D V C T H

T B X A O X Z N L

V C X C W N S O P

A Y T F Z Q D P C

L B S W F K X T N

Z N A O C Q P F Y

Clear-Sight Near Chart B

4

VISUAL FITNESS SKILL II: BIG-SIGHT

How "Big" Can You See?

Before we consider how "big" you can see, let's take another look at how "little" you can see. Can you constrict your vision enough to see one of those itsy-bitsy, teensy-weensy little letters at the bottom of the eye chart? As we've seen, this is the primary question that the routine eye exam asks. This is the reason that the doctor prescribes glasses and—no matter how perfect your glasses—the reason you'll need to exercise your Clear-Sight muscles.

Yes, how little can you see? Little enough to make out a single leaf on a tree, a blade of grass on the golf course, the tiniest letter on the medicine bottle?

Sometimes seeing *little* can even mean seeing *more*. The theatergoer who sees little can view the expressions on the actors' faces rather than just listening to costumed bodies moving about a stage. The pilot who sees little can spot a plane in the distance approaching at a thousand miles per hour.

Imagine having microscopic sight and being able to see the individual cells of your skin or the fibers in your clothing. Imagine being able to see molecules and atoms, electrons and quarks. Such science-fiction seeing would certainly expand your world and allow you to see more.

With seeing little in mind, let's rev up your Clear-Sight muscles one more time. Look across the room at some itsy-bitsy teensy-weensy little detail: if there's a picture on the wall, look at a detail in the picture. If there's a light switch, look at the switch or, better still, look at a single screw in the switch plate. Look hard at that detail for a moment, as hard and as focused as you can be.

Good. Now let me ask you a question. At the same time you were looking at the detail, how much of the rest of the room were you aware of out of the corner of your eyes? If you're not sure, repeat the experiment, looking as hard as you can at some tiny detail.

The truth is that sometimes seeing *little* isn't seeing *more.* Sometimes seeing "little" is just that: seeing little. The same commuter who has learned to block out his surroundings for the sake of reading on a train may have trouble retrieving his peripheral vision when driving at night. The same computer programmer who has learned to shrink the world to the size of a screen for many hours a day may have trouble expanding that world back to living size for the purpose of playing tennis or golf. For this reason, it's not enough to ask the question, How "little" can you see? We need to ask a second question: How "big" can you see?

How many words can you see at one time when you're reading? Are you aware of the position of the other players on the tennis court even when your racket's connecting with the ball? When you're driving, can you see the street signs and still be aware of the position of the cars around you? When you're golfing, can you open up your peripheral vision to read the green? When you're walking or dancing or otherwise moving, are you aware of your surroundings, or do you stumble or bump into things?

Just as seeing "little" has its uses, so does seeing "big." To answer *who* or *what* questions—Who is that person? What does that street sign say?—we depend on our ability to see "little." Answering *where* questions, however—Where is that ball? Where is that car? Where am I?—requires us to see "big." In chapter 1, we read about Darlene and her fear of night driving. Her inability to follow the road and judge the speed and position of other cars was caused by her failure to see big, not little.

Similarly, in chapter 1, Sarah's difficulty with reading had to do with a failure to see big. Sarah saw plenty little. It wasn't the littleness of the letters that confused her. It was seeing big enough to keep the letters and words in order. As a result, the words would slide around on the page. For example, the words *The cat* might blend into *Thc aet* or *Tcahet* or even *cat The*. At the same time, Sarah couldn't see big enough to predict where the words were going and where the sentences would end. No wonder she feared reading out loud to her class; she was never sure what letters or words might pop up next.

People like Sarah, who see the words running together or shifting or dancing on the page, are demonstrating difficulties with seeing big. If seeing big is difficult only within arm's reach, the problem will affect reading and computer work, but not driving or sports. If seeing big is difficult only beyond arm's reach, the problem will affect driving and sports but not reading or computer work. If seeing big is difficult at all distances, the problem will affect just about every area of life.

The general eye exam is designed to screen for actual physical problems that affect our ability to see little and big. There are health problems, for instance, that destroy central vision and make it impossible to see little. There are also health problems that cause "tunnel vision" and make it impossible to see big. Most of us, however, have no such physical problems. Instead, whether we see big or little or both depends on how we focus our attention when using our eyes. In actual fact, we use a different set of brain cells for seeing little than we use for seeing big. Which brain cells we choose to use, however, is a matter of visual fitness.

Our attention to the big and little points of life reflect not only how we see, but how we think. Albert Einstein once struggled with seeing little enough to perform basic arithmetic, but he later saw big enough for his calculations to change the way we view the universe. And how about Bill Gates? Would you imagine he loses himself in itsy-bitsy, teensy-weensy little details? Probably not: the ability to *think* "outside the box" requires the ability to *see* outside the box. And the ability to see outside the box requires the ability to see big. As far as visual fitness coaches are concerned, *seeing* big and *thinking* big can be one and the same thing.

As you might imagine, the ideal is to be able to see both big and little. Those who cannot make the shift from seeing big to little are likely to have trouble with attention. They will be distracted by everything around them rather than being able to zero in on what's in front of them. The child who's aware of every stray movement in the classroom but cannot stay focused during writing or reading assignments is an example of someone stuck in Big-Sight. Whether you're a child in a classroom or an adult in a lecture hall, you will need the ability to shift your attention to see as little as the task at hand requires.

Individuals' ability to shift from seeing big to little and little to big varies. Those in need of visual fitness workouts typically fall into one of two groups. The first group seems to be locked into seeing little. They understand the world around them by taking in one detail at a time and then using logic to add these details together. Such a process can be time-consuming—and if logic is flawed, the world becomes a confused place. The second group is generally locked into seeing big. Such individuals see the big picture but ignore the details. When members of the second group do summon the energy to see the details—as they may do from time to time—they lose the big picture.

For many activities, however, maximum efficiency requires us to see big and little at the same time. Darlene's driving and Sarah's reading are just two examples of how this ability to blend the big with the little can affect our lives. When driving at night, we must be able to see the dividing line and street signs at the same time that we're aware of the cars around us and the curve of the road. When reading (especially when learning) we have to be able to see the individual letters as well as words and sentences. The Big-Sight charts that follow are designed to build your ability to see not only big, but to see both big and little at the same time.

With this general information behind us, let's get started. Since everything you'll need to perform the Big-Sight exercises is right here in the book, you won't need to leave your seat. If you're reading through this book for the first time to get an overview of visual fitness, or if you're reading this book for information only, go ahead and take a few minutes to work through the first Big-Sight exercise to see for yourself what we mean by "seeing big." Then, if you'd like to, the next time you're driving or out for a walk, see how big you can see.

The Big-Sight Workout

Take a quick look at the six Big-Sight charts on pages 55, 58, 61, 63, 65, and 67. (You may want to photocopy them or cut them out of the book.) Each one is composed of a small picture or word in the center and increasingly larger pictures or letters toward the chart's outside.

As with the Clear-Sight exercises, follow your doctor's recommendations for wearing your glasses. If you have been told to wear your glasses for all near work—or all the time—wear them for these exercises. If your doctor has said that it's up to you whether or not to wear your glasses for reading, then take your pick, providing that you can see the details on the smiley face in Big-Sight I. If you wear bifocals that don't allow you to see the whole chart at the same time, you may have to do the best you can, either with or without your glasses, whichever is easier.

When you work with the Big-Sight exercises, leave both eyes open. Begin with Big-Sight I. So long as you're succeeding, work through each exercise quickly. If you run into trouble, however, back up to the last exercise or step you were able to work with easily. Spend more time on that earlier chart or step, then move forward, being sure you're mastering each step before progressing to the next. When you've reached Big-Sight VI, stick with that exercise in your visual fitness workouts.

With the exception of Big-Sight VI, you should hold the charts at your regular reading distance. If when you perform the exercises you have trouble with the pictures or letters running together and doubling up, these Big-Sight exercises are not for you. You may have an undiagnosed eye-muscle problem that requires the guidance of a professional visual fitness trainer.

Big-Sight I

In Big-Sight I there are seven steps. Master each step before moving on to the next. While doing the steps, try to remember to breathe and relax. Except on step 1, tightening your body and holding your breath will slow your progress.

1. Using the Big-Sight I chart (page 55), look at the smiley face in the center of the chart. Be aware of the details on the face. Try to see the black circles that make up the eyes. Try to see the two tiny circles of white that these black circles surround. Concentrate on these tiny eyes, their black and white, and see as "little" as you can see. Only on this step, tighten the muscles in your body and hold your breath.

2. Now, relax your seeing while you continue to look softly toward the smiley face. Out of the corners of your own eyes be aware of the eight gray squares. They form a rectangle. While you're doing this, don't stare hard at the smiley face or your vision will fade. To increase control of your peripheral vision, practice steps 3, 4, and 5.

3. Without taking your eyes off the smiley face, be aware of the three squares on the right. Now, still keeping your eyes on the face, shift your attention to the three squares on the left. Next, shift your attention to the three squares on the top, then the three on the bottom. Continue to shift your attention, combining the smiley face with the three squares on the right, left, top, or bottom until you can see the face and three squares at the same time easily. Remember to relax and keep breathing.

4. Now widen your attention until you can see the smiley face and the squares on both the right and the left at the same time. In other

Big-Sight I

words, see the face and the six squares that make up the two sides. This accomplished, shift your attention until you can see the face and the six squares that make up the top and bottom of the rectangle. Practice shifting your attention back and forth between the six squares making up the sides and those making up the top and bottom. Throughout these shifts in attention, keep your eyes pointing softly at the smiley face. Relax and keep breathing.

5. Once you can see the smiley face and six squares easily, open your side vision the rest of the way and practice seeing the face and all eight squares. Relax and keep breathing.

6. On this step you'll practice shifting between seeing "little" and seeing "big."

A. Begin by looking hard at the smiley face. Put your full attention on the white centers of the face's eyes. Work on making the face as clear as possible. Be aware of the relationship between the eyes and the smiling mouth. See all of the smiley face at once.
B. Relax your eyes. Keep them pointed at the face, but this time softly, not hard. Be aware of the face and the eight squares, all at the same time.
C. Continue alternating between steps A and B. See little, see big. See little, see big. Back and forth, back and forth. Relax and keep breathing. If the smiley face or any of the squares disappear, gently blink your eyes.

7. On this step, you'll practice seeing big and little at the same time.

A. Look softly toward the smiley face and be aware of the face and the eight squares at the same time.
B. Now focus in on the whites of the face's eyes while at the same time being aware of the eight squares.
C. Alternate between steps A and B until you can see the whites of the face's eyes easily at the same time you are aware of all eight squares. Again, if the smiley face or any of the squares

fade, gently blink your eyes. And remember to relax and keep breathing.

Big-Sight II

The chart for Big-Sight II (page 58) is the same as for Big-Sight I except that the rectangle of eight gray squares is surrounded by a larger rectangle of ten stars.

1. Look at the star in the upper right-hand corner. Be aware of that star and the two stars beside it. In other words, be aware of the three stars that make up the upper right-hand corner.

2. Move your eyes from corner to corner while being aware of the group of three stars that make up each corner.

3. Perform the following steps:

A. Direct your eyes to the center of the top row of stars and open your side vision enough to see the three stars that make up the top right-hand corner and the three stars that make up the top left-hand corner. In other words, see the pattern of six stars formed by the four top stars and the two side stars.
B. Shift your attention to the bottom six stars: the four that make up the bottom line and the two stars on the sides. In your mind, see big enough to combine these six stars into a single pattern.
C. Alternate your attention between steps A and B—back and forth, back and forth—until you can easily see either the top six stars or the bottom six stars. Throughout these steps, remember to relax and keep breathing.

4. Now perform these steps with the two sides of the star rectangle.

A. Shift your attention to the five stars that make up the right-hand side of the star rectangle. See these five stars as a group.

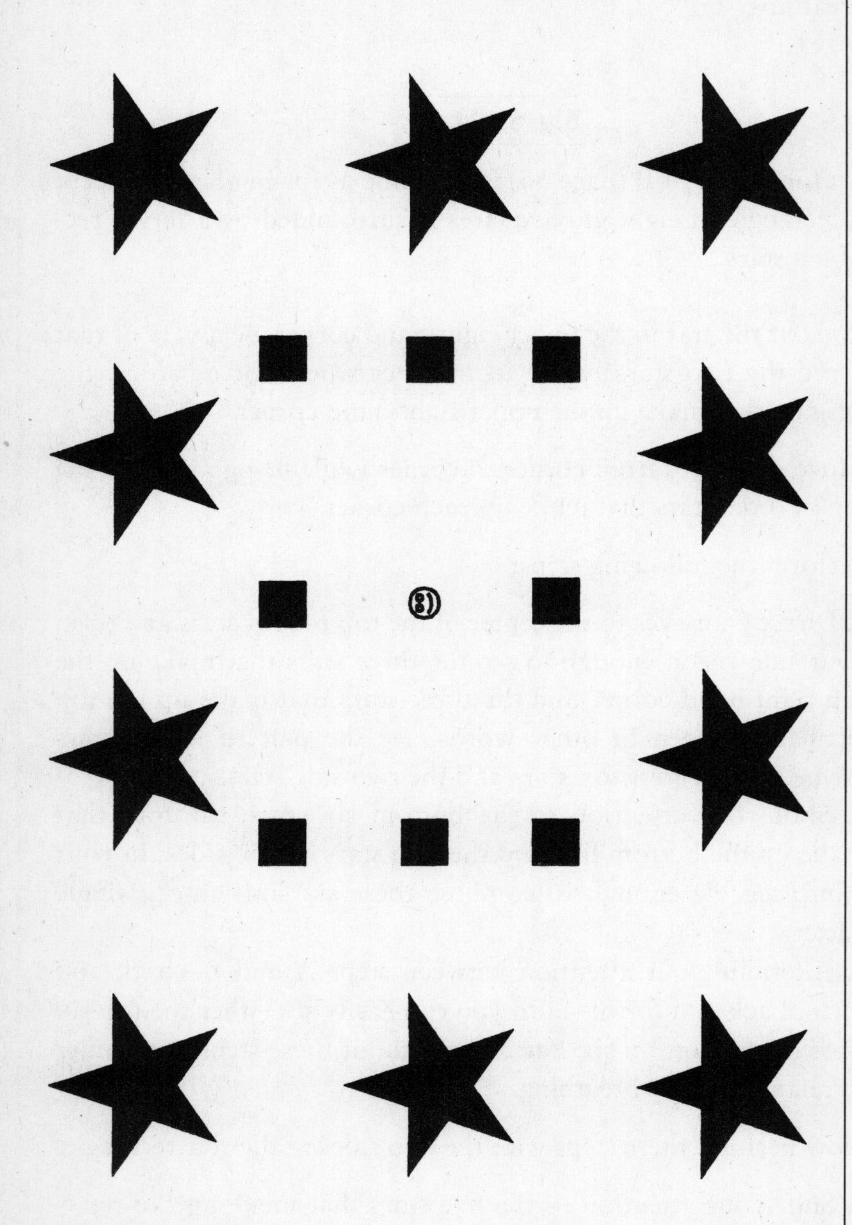

Big-Sight II

B. Shift your attention to the five stars that make up the left side of the star rectangle. See these five stars as a group.
C. Alternate between the right-side and left-side stars. Relax and keep breathing.

5. Perform these steps to see the entire star rectangle at the same time:

A. Alternate between seeing the top and bottom halves of the star rectangle.
B. Combine your view of the top and bottom halves of the star rectangle until you can see all ten stars as a group.
C. Alternate between seeing the right and left halves of the star rectangle. Relax and keep breathing.
D. Combine your view of the right and left sides of the star rectangle until you can see all ten stars as a group.
E. Rotate through these steps until you can effortlessly see the ten stars of the star rectangle as a group. Relax and keep breathing.

6. Now let's learn to see big and little at the same time.

A. As you learned to do in Big-Sight I, first see the whites of the smiley face's eyes at the same time you are seeing the eight gray squares. (If you have any problems return to Big-Sight I.)
B. Then shift your attention to see the ten black stars at the same time. (If you have any problems return to the earlier steps with this chart.)
C. Alternate between seeing 1) the smiley face and eight squares and 2) the ten stars, until you can do so effortlessly.
D. Now—remembering to look softly while relaxing and breathing—see the whites of the smiley face's eyes while simultaneously being aware of the eight squares and ten stars.
E. Continue rotating through these steps until you can effortlessly see the whites of the face's eyes while being aware of all eight squares and ten stars. If at any time the smiley face or any

of the squares or stars fade, gently blink your eyes. Throughout these steps, remember to relax and breathe.

Big-Sight III

The Big-Sight III chart (page 61) is the first of three that use letters instead of symbols. Though simple, the chart is designed to prepare you for Big-Sight IV. If possible, work through Big-Sight III quickly, then move forward to Big-Sight IV.

1. Look at the word *CLEAR* in the center of the chart. Make the word as clear as you can. Be aware of the white spaces inside the tops of the letters *A* and *R*. In others words, see little.

2. See big: relax your eyes and be aware of the word *CLEAR* and the eight letter *A*'s at the same time. On this step, don't worry about seeing the tiny details of the word *CLEAR*.

3. Alternate between these two steps until you can shift your attention effortlessly back and forth between seeing big and little.

4. Finally, see both big and little at the same time. Be aware of the white spaces in the tops of both *A* and *R* of the word *CLEAR* while at the same time seeing the eight letter *A*'s out of the corners of your eyes. If this is too difficult, continue to rotate through the first three steps. Then try again.

Big-Sight IV

The Big-Sight IV chart (page 63) is used to continue improving your ability to integrate your big and little seeing and prepare you for the charts that follow. Remember to relax and continue breathing while using the chart. If any of the letters fade, gently blink your eyes. If any of the steps seem too hard to do, spend more time on the earlier steps or earlier charts.

A A A

A CLEAR A

A A A

Big-Sight III

1. On this step, move your eyes wherever you wish. Find the rectangular arrangement of *B*'s between the *A*'s and *C*'s. As you did when you worked with Big-Sight II, practice shifting your attention through the following groups of *B*'s:

A. The three that form each corner of the *B* rectangle
B. The five that form the right side of the *B* rectangle
C. The five that form the left side of the *B* rectangle
D. The five that form the top of the *B* rectangle
E. The five that form the bottom of the *B* rectangle
F. All eight that form the complete rectangle

2. Continue rotating through the groupings in step 1 until you can easily see all eight *B*'s at the same time.

3. Now put your attention on the *C*'s. Repeat step 1 with these big *C*'s until you can easily see all eight *C*'s at the same time.

4. Practice shifting your attention between the *A*, *B*, and *C* rectangles until you can see any of the three.

5. Now practice shifting your attention to see more than one rectangle at a time:

A. The *A* and *B* rectangles
B. The *B* and *C* rectangles
C. The *A*, *B*, and *C* rectangles

6. Finally, with your eyes pointed toward the word *CLEAR*, shift your attention among the following:

A. See the word *CLEAR* combined with the eight *A*'s, as you did in Big-Sight III.
B. See the *B* and *C* rectangles at the same time, as you did in step 5 above.
C. See the entire chart at once.

When you have mastered the chart in this manner, you are ready to move on to the next chart.

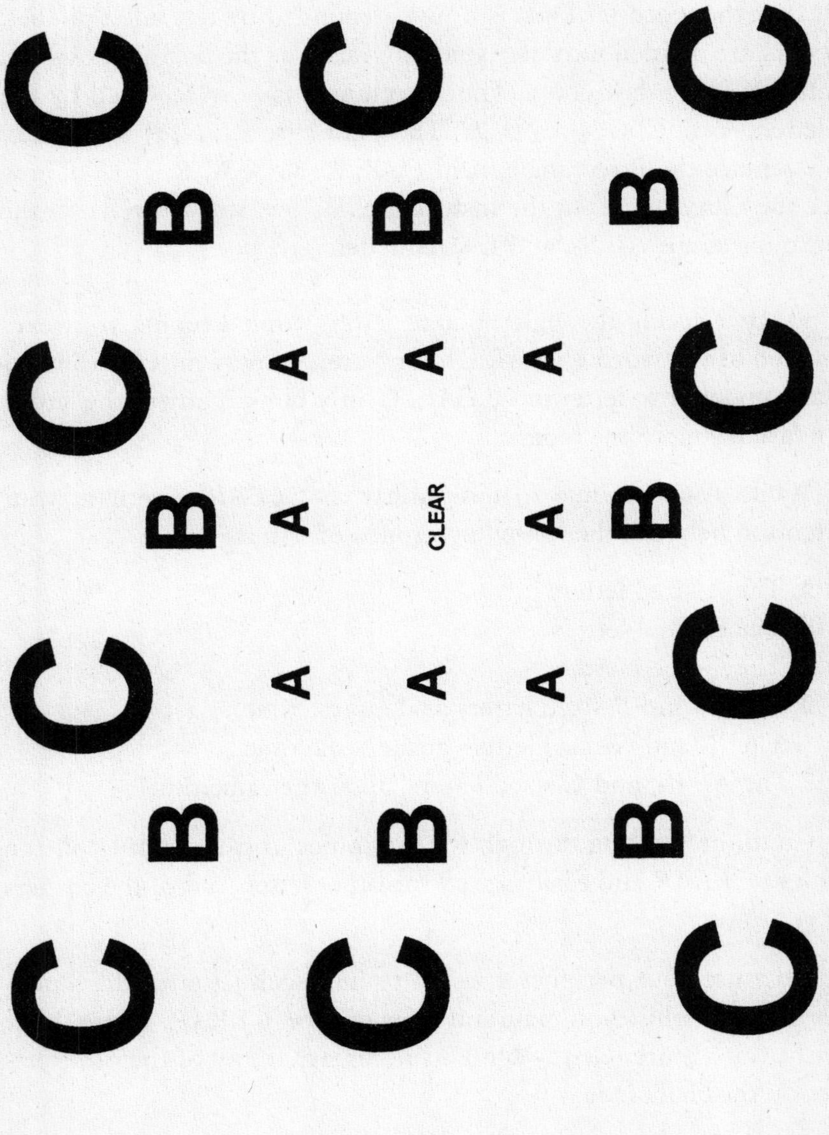

C C C C

B B B

A A A

C A B CLEAR A B C

A A A

B B B

C C C C

Big-Sight IV

Big-Sight V

The Big-Sight V chart (page 65) is much the same as the one before, except that the word *CLEAR* is now surrounded by the full alphabet. The letters are divided into the same three sizes as the rectangles of *A*'s, *B*'s, and *C*'s in the last chart. The *A* rectangle is now replaced by the eight letters *A, E, C, G, B, F, D, H*. The eight letters *I, N, P, K, J, O, L, M* now replace the *B* rectangle. And *Q, Z, X, V, S, W, R, Y, U, T* now replace the *C* rectangle. In the instructions below, we will refer to the three groups as the *A*-, *B*-, and *C*-sized letters.

1. Practice seeing the eight *A*-sized letters while keeping your eyes pointed at the word *CLEAR*. Don't stare. Allow your eyes to wander from letter to letter of *CLEAR*. Gently blink if any of the letters you are trying to see fade.

2. While you continue to look softly at *CLEAR,* alternate your attention between the following groups of letters:

A. The *A*-sized letters
B. The *B*-sized letters
C. The *C*-sized letters
D. The *A*- and *B*-sized letters at the same time
E. The *B*- and *C*-sized letters at the same time
F. The *A*-, *B*-, and *C*-sized letters, all at the same time

3. Continue rotating through the groupings in step 2 until you can look at *CLEAR* and easily see all the other letters out of the corners of your eyes.

4. Now use your peripheral vision to find each letter of the alphabet in order while you maintain your eyes on *CLEAR*. Again, look softly, don't stare. Gently blink as necessary if any of the other letters on the chart fade.

5. When you can perform step 4 easily, continue to find the letters in alphabetical order, but this time locate two letters at a time, put-

Q Z X V

I N P

A E C

T M H CLEAR G K S

D F B

L O J

U Y R W

Big-Sight V

ting your attention on *CLEAR-A-B*, then *CLEAR-B-C*, then *CLEAR-C-D*, and so forth through *CLEAR-Y-Z*. While you do this, also be aware of the rest of the chart.

6. When you can be aware of two letters at a time, find the alphabet three at a time—*CLEAR-A-B-C* and so on; then four—*CLEAR-A-B-C-D*. Expand your awareness to be as big as you can make it.

Big-Sight VI

The Big-Sight VI chart (page 67) is the same as the last one except that the letters appear in a less predictable order and the word *CLEAR* is printed in a smaller font size. Providing that you can see this smaller font and you've mastered Charts I through V, this is the chart you'll be using for your visual fitness workouts. If you cannot see the word *CLEAR* on this chart, then use Chart V.

When performing this exercise, you'll use the same steps you used with Chart V, but you'll add the following change: Instead of holding the chart at your normal reading distance, you'll practice holding it as far away from your eyes as you can while still being able to read the word *CLEAR* (providing your arms are long enough) and then as close to your eyes as you can while still making out the word *CLEAR*. For example, if you were trying to open your vision enough to see two letters at a time out of the corners of your eyes—as you did in Big-Sight V—you would perform the following steps:

1. Hold Chart VI at arm's length or as far away as you can while still being able to read the word *CLEAR*. Be aware of the whole chart while at the same time focusing on *CLEAR*, *A*, and *B*.

2. Move Chart VI as close as you can where you can still read the word *CLEAR*. Open up your side vision until you can see the entire chart out of the corners of your eyes. Focus on *CLEAR*, *B*, and *C* at the same time.

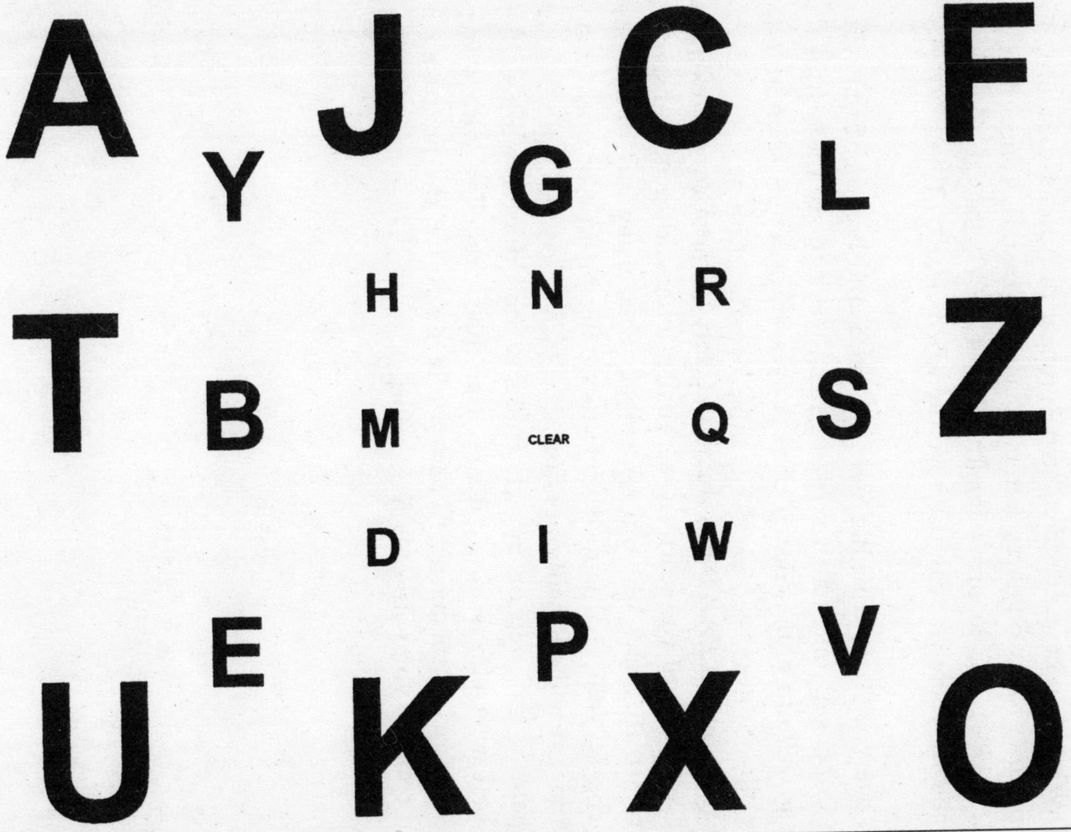

A J C F

Y G L

T H N R Z

B M CLEAR Q S

D I W V

E P

U K X O

Big-Sight VI

3. Repeat steps 1 and 2 until you have completed the alphabet: far—*CLEAR-A-B*; near—*CLEAR-B-C*; far—*CLEAR-C-D*; and so forth to *CLEAR-Y-Z*.

4. When you have mastered step 3, repeat the same procedure, but this time focus on three letters at a time; that is, concentrate on *CLEAR-A-B-C* at arm's length, then *CLEAR-B-C-D* as close as you can where you can still read the word clear *CLEAR*.

While you work with Big-Sight VI, relax and breathe. Even when you are focusing on smaller sets of letters, work to see the rest of the chart as well out of the corners of your eyes. If the letters seem to fade, gently blink your eyes.

If the word *CLEAR* is too small on Chart VI, perform this exercise with Chart V until you learn to see smaller letters at the near distance, then return to Chart VI. And remember, when using any of the Big-Sight charts, your goal is to be able to see big or little or both at the same time, at will and without effort.

Again, if any step on any chart seems too difficult, work earlier steps or earlier charts; don't waste your time struggling with a step you're not ready to master. When you're ready for the next step it should come to you with only mild effort.

5

VISUAL FITNESS SKILL III: DEEP-SIGHT

How Much Depth Can You See?

In the last chapter, we began to learn how to see "big," but there's more to seeing big than seeing simultaneously to the right, left, up, and down. Unlike the Big-Sight charts, life is not two-dimensional. There's also a matter of distance, a matter of depth.

Individuals with wandering eyes (of the physical rather than romantic variety) demonstrate reduced depth perception whenever their eyes wander. Even if your eyes are normal, however, there's no guarantee that your depth perception is all that it could be.

Doctors interested in thoroughness generally include a depth perception test in the routine eye examination. Sometimes they ask you to look into an instrument that resembles the one used by the Department of Motor Vehicles. Sometimes instead, they hand you a pair of black-framed sunglasses that could allow you to pass for one of the Blues Brothers. Whether you're looking in the instrument or wearing the glasses, they'll ask you to compare a number of symbols and determine which one is closest. The better your depth perception, the farther down the chart you'll be able to make your way.

These quick-to-administer depth perception tests, which are used during routine eye exams, are actually screening tests. Some of these tests reach only as far as "low-average." Some test up to "average." None tell us if our depth perception is above average. Thus, just as most people with normal eyes have better than 20/20 eyesight, most people with normal vision have better depth perception than will be assessed by the screening tests in doctors' offices.

There is another parallel between testing for eyesight and depth perception: depth perception tests, like eyesight tests, are centered around how "little," not how "big," we can see. Passing your depth perception test might suggest you can tell if one leaf on a tree is closer than a neighboring leaf. When you're judging the position of these leaves, however, there is no guarantee that you are also aware of how much space separates you from the tree; there is no guarantee that your Deep-Sight muscles aren't growing flabby.

Whenever you drive a car or catch a ball, you give your Deep-Sight muscles a workout and are rewarded if you can judge depth (or penalized if you can't). When you read or use a computer, however, there is no such reward for depth perception, no exercise for your Deep-Sight muscles. Whether you're lost in a book or in cyberspace, you're apt to lose awareness of how much real space separates you from the page or screen. All too often this lost awareness can cause you to hold the book too close or to stretch your neck toward the screen. As you adapt to this two-dimensional world of books and computers, you run the risk of giving up not only your Clear-Sight but your Deep-Sight.

Let's return to a demonstration I used in the Big-Sight chapter. Look across the room at a light switch and concentrate on some detail of that switch. Last time I assigned this task, I afterward asked you how aware you were of the rest of the room. This time I'll add, How much space were you aware of between yourself and the light switch?

There are a number of people who have no trouble reading for hours or working on a computer, but they've unknowingly given up their Deep-Sight. The physically flat world of books and computers provides us with pleasure, knowledge, and income, but as vision flattens to conform to the page or screen, we may experience headaches or sore eyes or need increasingly thicker glasses. Of greater concern, we may misjudge the position of cars on the road. Good visual fitness, especially exercising your Deep-Sight muscles, is therefore essential in reducing the visual pitfalls of prolonged reading and computer usage.

Before beginning the Deep-Sight exercises, let's perform two other demonstrations that will add to your understanding of the two-eyed nature of good depth perception.

1. Cover one eye and look across the room at some detail. While looking at this detail, be aware of how much space you can see between yourself and the detail.

2. Uncover your eye and be aware of how much more space you can now see. You should be able to see more "depth" with two eyes than with one.

To understand more about how your two eyes add together to improve your depth perception try this second demonstration:

1. Hold your left hand up about a foot in front of your eyes. Spread your fingers. Position your hand so that your thumb is closest to you, but turn your hand enough so you can see the spaces between your spread fingers.

2. While you maintain your left hand in this position, cover your right eye with your other hand. Be aware of how much space you can see between the fingers of your left hand.

3. Move your right hand over to cover your left eye. Again, be aware of how much space you see between the fingers of your left hand.

4. Repeat steps 2 and 3 several times. Be aware of how each eye has a slightly different view of your raised hand, and how one eye can see more space between the fingers than the other eye can.

5. Finally, leave your left hand in the same position, but uncover both eyes. See how your two different views of the hand combine into a third view, which is more three dimensional than the view from either eye alone. Be aware of how much more space you can see between your fingers. Be aware of how much more space you can see between your face and the hand.

This is how two-eyed depth perception (also called *stereopsis*) works. At closer distances, at least, each eye enjoys a different view. When both eyes are open, the Deep-Sight muscles combine these two views into a single, more three-dimensional view. Such three-dimensional seeing adds to our ability to drive, read a green on a golf course, or return a lobbed tennis ball. And, as we will learn in the exercises that follow, the key to strengthening the Deep-Sight muscles is, once again, seeing big.

In addition to improving your awareness of depth, the following exercises may also help you to relax your eyes and reduce your eyestrain and headaches and general tension. Since all three Deep-Sight exercises require no charts or other equipment, you can perform them anytime you feel your eyes beginning to strain or your head beginning to ache.

The Deep-Sight workout is brief. If you've ever felt as if the computer bytes are chewing up your eyes or forehead, give Deep-Sight I, II, and III a whirl. Even if you're reading this book purely for information and have no interest in becoming a visual athlete, take a few minutes to learn these exercises. Your effort will be repaid in full.

Your Deep-Sight Workout

Deep-Sight I

Stand up and walk to the center of the room. Turn and face a wall. Move forward or backward until you find a position close enough to the wall to force your eyes to stretch when, without moving your head, you look to the wall's four corners. The closer you stand to the wall the more your eyes will have to stretch, so select a distance that is comfortable and allows a gentle stretch, not a strain.

Stand with your feet spread to shoulder width. Pretend that the four corners of the wall are the bases of a baseball diamond. Down and to your right is home plate. Up and to your right is first base. Up and to your left is second base. And down and to your left is third base.

You are going to run the bases with your eyes. To begin with, direct your eyes from base to base. Home, first, second, third, home. Pause at each base.

Without moving your head, make a few rounds of the bases. Then run the bases in reverse order: home, third, second, first, and home.

When you first begin, you may feel your eyes and stomach begin to protest. If so, you're too close to the wall. Back up until you're comfortable. As your eyes grow accustomed to the exercise, you can gradually work your way closer to the wall.

Deep-Sight II

Now that you've made a few rounds of the bases to get the hang of the procedure, add the following modifications:

1. Extend one arm out in front of you until your hand becomes the "pitcher's mound" in the middle of the "baseball diamond."

2. As you "run the bases" open up your side vision, stop at each base, and be aware of your hand and as many bases as possible at the same time out of the corners of your eyes. Also be aware of how much space there is between yourself and the wall, how much space there is between your hand and the bases.

Spend as much time on Deep-Sight II as it takes until you can run the bases without losing awareness of the amount of space between you, your hand, and the wall.

Deep-Sight III

Once you can run the bases while keeping your side vision open, there's one thing left to add: your breathing pattern. Persons performing eye exercises (especially those who are nearsighted) tend to hold their breath. Don't. Whatever method you use to remember to breathe is fine. One way is to inhale as you move to one base, and then exhale as you move to the next base.

For Deep-Sight III, see if you can run the bases and keep your side vision open while remembering to breathe. If not, go back to Deep-Sight II and practice it some more before returning to Deep-Sight III.

Deep-Sight III is designed to relax your eye muscles and open up your peripheral vision and depth perception. After you work the procedure, you should be more aware of depth, more aware of the space between objects.

Once you've mastered Deep-Sight III, you'll include it in your visual fitness workouts. You should also use the exercise at the first sign of any eyestrain or headaches during your desk or computer work. It's better to relax your eyes at once rather than wait until you have a full-blown headache.

As you master Deep-Sight and become more aware of the space between yourself and where you're looking, add this awareness to your daily life. Whether you're walking, jogging, or driving, try to see how much space you can be aware of, how "deep" your vision is. When you're using a computer and you find your neck and head stretching forward, remember there's an actual room around the computer screen and that you're sitting in that room. Be aware of the space between yourself and the computer screen and remember to breathe. The same is true when you find your book creeping ever closer to your face: relax, breathe, and be aware of how much space separates you from the pages. Open up your side vision for a moment and be aware of the room around you. See how big and deep you can make your vision.

You can also rest your eyes by looking above the book or computer screen and seeing into the distance. While doing this, pretend that you're a fish and the world is a fish bowl. Try to open your vision big and deep enough to see all the "water" around you at the same time.

There's one more advantage to Deep-Sight. As mentioned in previous chapters, many of our problems are not in front of us; we carry those problems in our minds. Next time you're feeling blue or anxious, spend some time on Deep-Sight III and see if you don't also improve your point of view.

6

VISUAL FITNESS SKILL IV: STRONG-SIGHT

How Long Can You See?

Okay, so you now have the tools to make your vision clearer, bigger, and deeper. But how long can you see? By *long* we're not talking about distance; we're talking about time. How long can you see before you begin to tire out? Do you—like Sarah in chapter 1—count a book's pages before you consider reading it? And, when you do find a book that's thin enough, do you fatigue rapidly and lose comprehension or fall asleep? When you play tennis, do you have to fight to maintain concentration during later sets? When you're driving for longer distances, do you get headaches or want to fall asleep? Does your desk or computer work grow more frustrating as the day progresses, and—like Todd in chapter 1—by the time you get home are you too exhausted to enjoy your friends or family?

If you answered yes to any of these questions, then it's time to add the Strong-Sight exercises to your visual fitness workouts.

Actually, all the exercises we've examined so far can help you to see longer. Especially important, however, are the Clear-Sight and Strong-Sight exercises, because in addition to working our imaginary set of mental muscles, these exercises work to coordinate the actual, physical muscles that focus and team your eyes.

Indeed, each eye has seven physical muscles. As I described in the Clear-Sight chapter, there's one muscle inside each eye, which we use to keep things clear. In addition, there are six outside-the-eye muscles, which we use to point or aim the eye. If we consider both eyes, the number of muscles that we have to coordinate increases to fourteen, and this coordination becomes even more complex when we consider the system of nerves that are required to make all this happen.

The nerves of the body are run mainly by two systems, one involuntary, the other voluntary. The involuntary system speeds up our heart rates when we get excited, sends juices to our stomachs after we've eaten, and shuts down those juices when we're in danger of being eaten ourselves. The system is called involuntary because, without considerable practice, we have little control over it.

The voluntary system, on the other hand, is under our control. We can use it at will to wiggle our fingers and toes, our arms and legs.

Wiggling our eyes, however, provides more of a challenge because both the involuntary and the voluntary systems are involved. The Clear-Sight muscles inside the eyes are controlled by the involuntary system. We look at something and it clears automatically—or it doesn't. The muscles on the outsides of our eyes are voluntary. Within certain physical limits, we can point our eyes wherever we want to. To make things even more complex, the muscles outside our eyes have approximately ten times as many nerves controlling a given number of muscle fibers as do the leg or neck muscles. While one nerve might control one hundred muscle fibers in a leg muscle, a single nerve might control ten fibers in an eye muscle. As a result, the muscles that move our eyes are some of the most accurate in the body.

Even so, when we put all these facts together, they all too often spell trouble—especially when we're performing the unnatural act of reading.

"Unnatural?" you ask.

Yes. Unnatural.

Not too long ago, humans were more interested in hunting—and in not being hunted—than in reading. Then came farming—another task beyond arm's reach. It wasn't until the fifteenth century that Gutenberg invented movable type and started the trend toward affordable reading materials, and even so, books remained luxury items until relatively recently.

Yes, reading is an unnatural act. Not only do we have to coordinate fourteen eye muscles, we have to coordinate those muscles through two very different systems of nerves: not an easy task, especially when we're under stress.

As mentioned above, when things get dangerous, your stomach shuts down. When you're about to be eaten by a saber-toothed tiger, for instance, you need all your energy for fleeing, not digesting your Wheaties. Similarly, when that tiger's breath is warming the back of your neck, it's not a good time to be studying your *TV Guide*. As a result, when your body gears up to get out of Dodge, it shuts down not only your digestion, it shuts down your eyes, at least for seeing clearly up close.

Fortunately, there aren't a lot of saber-toothed tigers roaming the streets these days. Unfortunately, as far as your body is concerned, stress is stress. A deadline at work can be every bit as alarming as a hungry tiger. Whether you're a computer programmer restraining your body unnaturally in a seat as you compete for a promotion at work, or a child restricted from normal movement as you compete for grades at school, your body is likely to sense the stress and go into tiger-fleeing mode. As a result, the same system of nerves that speeds up your heart rate can cause the focusing muscles inside your eyes to relax. To counteract this relaxation, we have to use extra effort to keep the reading material clear. This extra effort produces more stress and the increased stress requires more effort until it all spills over into the way our two eyes coordinate together. All this commotion creates havoc with our visual fitness. If you have any doubt, compare your tennis game as it is after a day of desk or computer work to how it is on weekends when your eyes are relaxed.

Now, none of this is to say that all reading reduces our visual fitness. It's the combination of reading and stress that creates the problems. A person who can effortlessly read at a college level is unlikely to encounter any visual stress when reading a bestselling novel written at a sixth grade level. It's not until you approach the edge of what you can understand that the stress—and the breakdown of visual fitness—begins. Learning to read, tackling a new computer language, becoming acquainted with the latest updates in technology can all be stressful, all hard on your visual fitness and even your eyes. But once visual fitness begins to dip, then all reading—even ploughing through those reader-friendly bestsellers—can become a challenge, which in turn leads to more stress and more breakdown in visual fitness.

A vicious cycle.

According to visual fitness pioneer Dr. A.M. Skeffington, whom we'll learn more about in chapter 11, we respond to this combination of stress and reading in one of four ways:

1. We avoid reading and desk work and go into sales.

2. We continue reading and learn to put up with the eyestrain, fatigue, and inefficiency.

3. We unconsciously adapt our eyes for near seeing and give up our distance vision.*

4. We unconsciously eliminate the fight to team our eyes by learning to rely on one eye instead of two (with a resultant reduction in eye-hand coordination and depth perception).

Fortunately, with this book in hand, you now have another choice: you can increase your visual fitness to meet the demands of learning and earning and fun. You can reduce your premature fatigue when

---

*Close to 80 percent of Ph.D.s—who evidently have mastered the world within arm's reach—have given up their ability to see far away without glasses. Whether these Ph.D.s were genetically predetermined to be nearsighted scholars or whether the use of their eyes caused their nearsightedness remains a bone of contention among vision scientists. My own view is that a combination of genetic and environmental factors affect the development of nearsightedness, at least in children. In the case of adults who become nearsighted only after they begin spending hours in front of computer screens, I lean more toward Skeffington's model.

using your eyes for reading, driving, or sports and also reduce the number of days you leave work too tired to enjoy your loved ones or too annoyed to tolerate anyone in your space.

If this is your first time through the book, or you're reading for information rather than to solve a visual fitness problem, spend a little time on the first seven steps of the exercise below. The effort will give you a little eyes-on experience to improve your understanding of what we mean by Strong-Sight. Once you're ready to begin your visual fitness program, you'll add these exercises to your workout.

Your Strong-Sight Workout

The following exercises are designed to increase your visual fitness, not to treat any medically significant eye-muscle problem. If you have strabismus (an eye that wanders in toward your nose or out toward your eye) or if you've had surgery for strabismus in the past, do not do these exercises. You'll need a doctor to guide you. (You'll learn more about finding such a doctor in chapter 11 and Appendix A.) The same is true if your doctor has told you that you have a medically significant eye-muscle problem making it difficult to use your eyes within arm's reach (convergence insufficiency or convergence excess).

Strong-Sight I

This exercise is designed to allow you to improve your depth perception and at the same time learn to point your eyes together comfortably within arm's reach. Strong-Sight I makes use of the two-eyed depth perception we learned about in the Deep-Sight chapter. This nine-step exercise begins with five introductory steps:

1. If you are right-handed, use your left hand to hold this book upright, its front cover facing you, at your normal reading distance. Slip your left thumb between the front cover and the rest of the book so the cover is a little closer to you than the pages.

2. Use your right hand to hold a pen two or three inches to the right of the book, with its tip pointing left, toward the upper right-hand corner of the book's cover. (If you are left-handed, hold the book, its back cover facing you, with your right hand and the pen with your left.)

3. Once you believe you have the pen tip lined up with the cover's corner, move the pen slowly to the left until the tip of the pen touches the corner of the book. You want the pen to remain parallel to your face and at the same distance away from you as the book. In other words, don't move the pen up close to your face and point its tip out toward the book so that you can use one eye to line it up with the book's corner as if you were sighting a gun. If you did the procedure in this manner, it would not force you to use both eyes together.

4. (Optional). Just for fun, to understand better how this drill is forcing you to use both eyes, close one eye—if you can—and repeat the pen placement. Can you see how hard it is to line up the pen when using only one eye?

5. If, with both eyes open, you miss the book's corner with the tip of the pen on your first attempt, do not try again until you have moved the pen back to its starting position two or three inches to the side of the book. At this point, try to see big: be aware of both sides of the room at the same time. Again, line up the pen tip with the book's corner and move the pen until its exact tip touches the exact corner of the cover. As you perform the placement, remember your Deep-Sight exercises and try to be aware of how much space there is between your eyes and the book. In other words, try to see not only big, but deep.

6. Once you've mastered touching the corner of the cover with the tip of the pen, you're ready to begin the actual Strong-Sight exercise, which has three parts:

A. Hold the book and your pen about ten inches in front of your nose. With the pen to the side of the book, make the same alignment you learned above. Then move the pen toward the book and touch the cover's corner with the tip of the pen. Relax, breathe, and see big as well as deep.

B. Hold the book and pen at arm's length. Repeat the pen movement.

C. Lower the book and look across the room at the center of a wall. Be aware of the wall's four corners with your side vision, and while you relax and breathe, see how much space there appears to be between your face and the wall.

Repeat the three parts of step 6 until you can effortlessly make the pen placement at both the ten-inch and the arm's-length distances. If the ten-inch distance hurts your eyes, move the book out to twelve inches and repeat steps A through C. Remember to keep seeing big as well as deep.

Seeing big is especially important if you're having difficulty at arm's reach after working at the closer distance. If this is the case for you, do not leave step 6 until you can effortlessly do the pen placement at arm's length after working at the nearer distance. When you do make the pen placement at arm's length, open up your side vision. Relax and breathe and be aware of as much space as you can just as you're learning to do in the Deep-Sight exercise.

If you are approaching (or retreating from) age forty, your vision is likely to be blurry when holding the book at ten inches. Ignore the blur and concentrate instead on your pen placement and on being aware of how much space there is between you and the cover corner.

If your pen starts missing the cover's corner at closer distances, it is likely that you are not using both eyes at these distances. Think about crossing your eyes and looking out of both eyes at the same time when performing the near placement. If you continue to have difficulty judging the position of the corner, move the book as far away from your face as you need to in order to make an accurate pen placement. Then, as you improve, gradually work the book back closer to your face.

If, when you're doing the near-seeing step of this exercise, the book or the pen seems to be doubling up, hold the book far enough away from your eyes so that no doubling occurs.

7. When you've mastered step 6, begin holding the book at eight inches when doing the near-seeing step. With the book at this eight-inch distance, repeat the three parts of step 6 (near, arm's length, and distance) until they are effortless. Everything that applied to steps 5 and 6 applies here.

8. Next, repeat the near, arm's length, and distance steps while you hold the book at six inches for the near step.

9. Finally, perform the three steps while you hold the book at four inches for the near step.

If the drill gives you a headache, and if it's the type of headache that normally bothers you, good! The same exercise that's causing your headache will eventually cure your headache. True, you may have to take more breaks, or perform the near placement at twelve or more inches before moving to ten, eight, six, and four. You may have to begin with two minutes a day on this exercise rather than seven, or you may have to spend more time on looking across the room and being aware of how much space you can see. Whatever you have to do, however, don't despair. Just continue to breathe, relax, and see big while performing Strong-Sight I. (As mentioned in chapter 2, if you have persistent headaches, don't forget to get them checked out by a physician.)

As for how much time to spend on each step before progressing to the next, it will take as much time as it takes before you are comfortable with a step. For some, that might mean one session. For others that might mean weeks. Don't try to rush your Strong-Sight exercises. On your first day, content yourself to alternating between ten inches, arm's length, and the wall—even if you're not particularly uncomfortable. If, after your workout is over, your eyes aren't sore, then you're ready to move closer during the next session.

The first few sessions, the Strong-Sight exercises may wear you out. You may experience the same exhaustion you feel after a day at work.

Again, this is good. Proceed slowly. Spend a number of sessions at the easier distances before moving the book closer. You'll probably find that within a half dozen sessions you will already be doing better.

When you have mastered Strong-Sight I and can easily and accurately touch the pen tip to the book cover's corner when alternating between four inches and arm's length, then, and only then, are you ready to proceed to Strong-Sight II.

Strong-Sight II

Most of us take it for granted that if we have both eyes open, then we are using both eyes. This assumption is not always true. Many people who have normal eye exams nevertheless ignore or suppress some of the information coming in through one eye or the other, or sometimes both. Strong-Sight II will allow you to use more of the information coming in through both eyes.

To understand these remarks, consider the Strong-Sight Line (on page 85). Based on a procedure popularized by Staten Island eye doctor Frederick Brock in the 1940s and '50s, the chart is made up of a single line with a dot in the middle of it.

With your book open to the chart and that page facing the ceiling, hold the book parallel to the floor and a few inches beneath your eye level. Move the book until the closest end of the page is about six inches in front of your nose and the dot on the line is about a foot from your nose. Now adjust the position of the book until the line is dead center between your two eyes. With the book in this position, look down the line at the dot and gently blink your eyes.

If both your eyes are aimed exactly at the dot, and if you are not ignoring information from either eye, the single black line will seem to divide into two black lines, which form an X crossing at the dot. You may have to blink your eyes a few times before the second line appears. Here's how these two lines should look. Your right eye sees line A—the line that starts to your left and slants toward your right as it moves farther away from you. Your left eye sees line B—the line that starts to

Strong-Sight Line

your right and slants toward your left as it moves farther away from you. If line A seems to fade in and out, it means that you are ignoring information coming in through your right eye. If line B seems to fade, you're ignoring information from your left eye. If lines A and B fade in and out in turn, you're ignoring information alternately from both eyes.

View of Strong-Sight Line When Eyes Point Directly at Dot

To confirm for yourself which eye sees which line, continue watching the dot and cover one eye at a time to see which line disappears.

Not only does the Strong-Sight Line tell you if you're ignoring information from your eyes, it tells you how accurately you are aiming your eyes.

If when you look at the dot, the lines cross behind the dot, as in the picture on the next page, then your eyes are pointing behind the dot. To correct the position of your eyes and get the two lines to cross at the dot, you will have to think about tightening your eyes and aiming them at a point on a portion of the line that is closer to you than the dot.

If when you look at the dot the lines cross in front of it, as in the picture on page 88, then your eyes are pointing in front of the dot. To correct the position of your eyes and get the two lines to cross at the dot, you will have to relax your eyes and think about looking farther away.

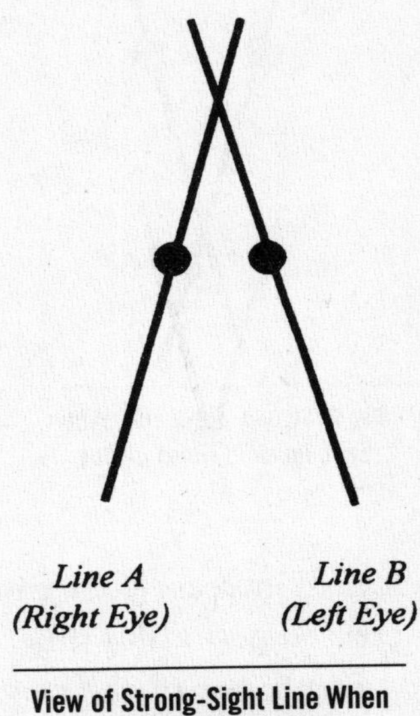

View of Strong-Sight Line When Eyes Point behind Dot

If when you look at the dot you continue to see only one line, you are using only one eye. Should this occur, look at the dot for a minute or two and gently blink your eyes. Keep your eyes on the dot, but think about seeing the entire room out of the corners of your eyes. Try moving the book a little closer or farther away. Try raising or lowering the book an inch or two or moving the book a little to the right or left. You'll probably begin to see two lines.

Should you continue to see only one line, then—for whatever reason—you've learned to ignore information coming in through one of your eyes. If so, this exercise and the Strong-Sight exercises that follow are not for you, at least until you see a visual fitness trainer. In the meantime, continue using Strong-Sight I during your visual fitness workouts. Even if you're unable to do Strong-Sight II through V,

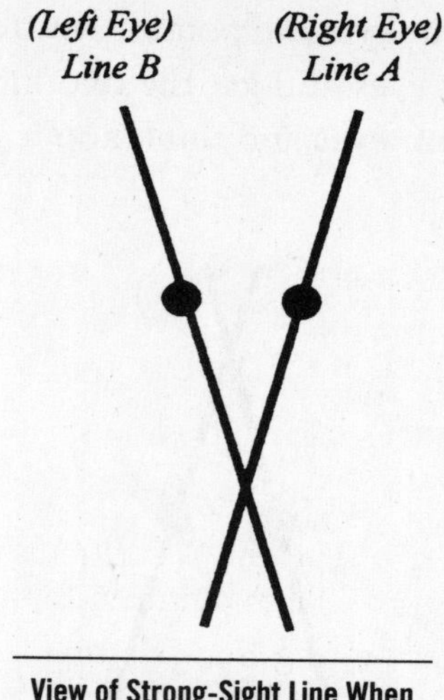

View of Strong-Sight Line When Eyes Point in Front of Dot

Strong-Sight I, all by itself, will increase your endurance and allow you to see comfortably for increased periods of time.

These preliminary comments out of the way, we're ready to begin the steps of Strong-Sight II:

1. Using the Strong-Sight Line (on page 85), position the book as explained above so that you're looking down the line toward the dot, which is about a foot from your eyes.

2. Be aware if one or both of the lines are fading in and out. If line A is fading, be aware of more of the right side of the room out of the corner of your right eye. If this action doesn't make line A more solid, then use a finger of your free hand to touch your temple next to your right eye. Imagine moving over inside your head toward that temple. Imagine looking out of your head from this new, right-side position. This shift in viewpoint should strengthen line A.

If line B is fading, try to be aware of the left side of the room out of your left eye. Touch your left temple with your free hand and imagine yourself sliding over inside your head and looking out of your left eye.

3. Once you have strengthened the fading image, touch your other temple and shift over inside your head as if you're looking out of your other eye. Try to make the weak image fade. Then shift back until you're looking out of your weaker eye and make the image come back. For instance, if line A was the one that fades, you'd start by touching your right temple and looking harder out of your right eye until the image of line A became stronger. You'd then touch your left temple and pretend you were shifting over to look out of your left eye until line A again faded. You'd work back and forth, back and forth making line A strengthen and fade until you were in complete control of both lines.

4. If instead of one line fading, both lines are alternately fading, perform the following sequence:

A. Open up your side vision to both sides at the same time and see as "big" as you can. In other words, while continuing to look at the dot, be aware of the entire room out of the corners of your eyes. Relax and breathe.
B. Once you've gotten both lines to become bolder, look hard at the dot and see as "little" as you can. Forget about your peripheral vision, and see if you can make the lines fade again.
C. Relax and be aware of the periphery until the lines are once again bolder.
D. Continue alternating between seeing big and seeing little until you can control the boldness of the lines.

5. Once you've learned to keep both lines from fading, you can begin to work on the accuracy of pointing your eyes directly at the dot. If the lines cross behind the dot, tighten your eyes and think about looking nearer to you on the page until the two lines cross at the dot. If the lines cross in front of the dot, relax your eyes and think about looking farther away until you can make the lines cross at the dot. If you are having persistent difficulty making the lines cross at the dot, return to Big-Sight I for more work on controlling your eyes to point closer and farther.

6. Once you've learned to make the lines cross at the dot, try the following sequence:

A. Tighten your eyes and make the lines cross a few inches in front of the dot.
B. Relax your eyes and make the lines cross a few inches behind the dot.
C. Position your eyes until the lines cross exactly at the dot.

Repeat this sequence—incorporating what you learned in steps 2 through 4—until you can easily make the lines cross at the dot without either line fading.

7. These steps achieved, perform the following sequence:

A. Move the edge of the book until it touches your nose and repeat step 6.
B. Move the book until its edge is about a foot from your face and repeat step 6.

Repeat this sequence until you can voluntarily control the position where the lines cross at both distances.

Strong-Sight III

After you've mastered Strong-Sight II and can voluntarily control the position of your eyes, you're ready for Strong-Sight III. With this exercise you'll continue to increase your endurance and concentration during visual tasks.

Turn to the Strong-Sight III chart (on page 91). This chart is made up of two solid circles almost touching one another. If you can't cross your eyes, begin with step 1. If, after your work with Strong-Sight I and II, you can voluntarily cross your eyes, begin with step 2.

Strong-Sight III

1. This step is for those who followed the advice, Never cross your eyes or they'll stick. If indeed you never learned to cross your eyes, you'll need a pencil to help you with this exercise:

A. Hold the Strong-Sight III chart at about reading distance and the pencil in front of it as shown here:

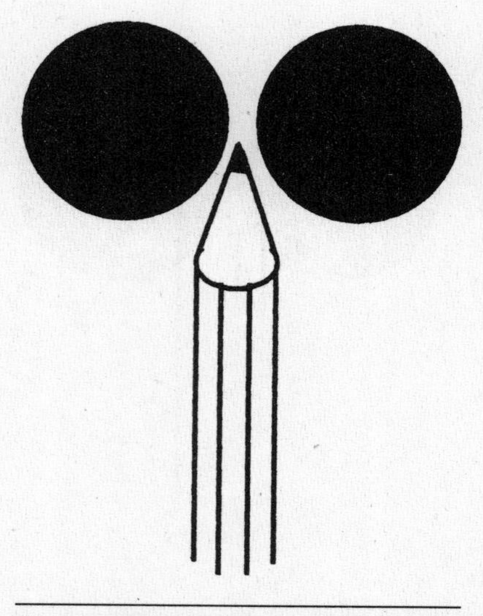

Pencil Held in Front of Strong-Sight III Chart

B. Keep looking at the tip of the pencil while you move it toward you, and with your peripheral vision, be aware of the circles until they become three as shown on page 93.

C. Keep your eyes crossed at this distance and concentrate on the center circle while you move the pencil down and out of the way. Providing that you keep your eyes slightly crossed and don't move them back out to the page, you will continue to have three circles. If this step is too difficult, then you may need more work on Strong-Sight I and II. For the time being, however, move on to step 3.

2. Hold the Strong-Sight III chart at about reading distance and cross your eyes just far enough to make the two circles move into three as shown below. If you cross your eyes too far, the circles will double into four circles. Should this occur, relax your eyes just enough to get three circles. So long as you see three circles, you are not shutting off either eye.

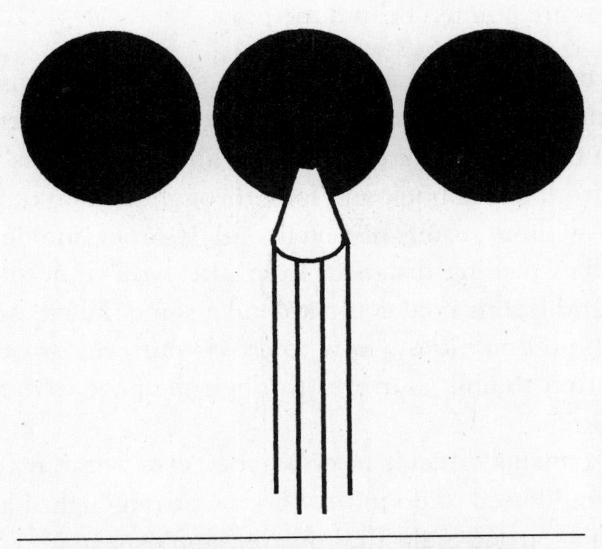

Pencil Moved toward Your Eyes until You Get Three Circles

How Strong-Sight III Chart Appears When You Perceive It as Three Circles

3. After you have attempted steps 1 or 2, move the book toward your face until the Strong-Sight III chart that is only a few inches in front of your nose. Relax your eyes and, as if you were Superman or Superwoman, stare through the page until the two circles become three. If instead you get four circles, try moving the page a little farther away until the circles merge into three. Once you have three circles, concentrate on the center circle and slowly move the book farther away from you while you maintain three circles. Your goal here is to be able to maintain three circles at reading distance while your eyes are pointed behind the page.

4. Alternate back and forth between steps 2 and 3 until you can easily make three circles by either crossing your eyes or looking through the page. Your goal is to be able to alternately cross your eyes in front of the book and look through the book to make three circles—without using a pencil and without moving the book closer than reading distance. Even after you've mastered Strong-Sight I and II, this goal could still take some time to achieve, especially if you don't know how to cross your eyes or can't use your imagination to aim your eyes through the page to a point behind the book.

If it remains difficult to cross your eyes without the help of a pencil, you'll need to do more work on Strong-Sight I and II. While performing Strong-Sight III, however, don't spend all your time trying to cross your eyes. After each attempt at crossing your eyes, make an attempt at looking through the page.

If it remains difficult to look through the page, substitute two pennies for the circles on the Strong-Sight III chart. Hold the bottom tip of one penny between the thumb and forefinger of one hand. Hold another penny between the thumb and forefinger of your other hand. Position the pennies side by side just as the two circles are positioned on the chart. Hold the pennies about a foot in front of your face and look through them across the room until you can turn them into three.

5. Once you can cross your eyes to get three circles without a pointer, you can use the two pennies instead of the chart:

A. Look through the pennies and make three.
B. Cross your eyes and make three.

C. As you improve at A and B with the pennies almost touching side by side, begin to increase the separation of the pennies until you can do the exercise—eyes crossed and eyes far—with the pennies held about an inch to an inch and a half apart.

Strong-Sight IV

When you have mastered Strong-Sight III using two pennies, without a pointer, move on to the Strong-Sight IV chart (page 96).* This chart adds the perception of depth to Strong-Sight III. Perform the following steps:

1. Hold the Strong-Sight IV chart about ten inches in front of your eyes. Look through the two sets of ovals until you have three sets of ovals. In the middle set, the smaller, center oval should appear to pop out closer than the larger oval. If the center oval appears farther away than the larger oval, you are crossing your eyes, not looking far away. Should this occur, try holding the chart a little closer and practice staring through it.

2. Cross your eyes and make three sets of ovals. In the middle set, the smaller, center oval should appear to be farther away than the larger oval. If the center oval is floating toward you, you're looking through the chart, not crossing your eyes. Use the pencil, as you may have used it in Strong-Sight III, to cross your eyes.

3. Go back and forth between steps 1 and 2 until you can easily make the center oval pop alternately near and far. If crossing your eyes is difficult, continue to go back and forth anyway. If looking through the ovals is difficult but crossing your eyes is easy, spend most of your seven minutes practicing looking through the oval chart.

---

*The opaque chart in this book is difficult to look through. If you photocopy this figure on plastic, you'll have an easier time. If you'd prefer, you can contact the author and he'll send you plastic targets for both Strong-Sight IV and V. You'll find the author's e-mail and snail-mail addresses in Appendix A.

Strong-Sight IV

4. When you can easily perform step 3, adjust your eyes until you get three sets of ovals, then practice rotating the Strong-Sight IV chart while you maintain three sets with the middle set three-dimensional. If crossing your eyes is easy and looking through the chart is more troublesome, spend most of your time rotating the chart while you are looking through it and the center oval is popping toward you.

Strong-Sight V

When you've mastered Strong-Sight IV, you're ready for Strong-Sight V.

Begin with the Strong-Sight V chart (on page 98) the same way you did with Strong-Sight IV:

1. Pretend you have x-ray vision and look through the page to get three sets of ovals. In the middle set, get the three ovals to pop out toward you. The solid center oval should be closest, then the medium-size oval. The largest oval should be farthest away. It may help to jiggle the chart to get the three ovals to pop out as much as possible. If the three sets of ovals are partly overlapping, then you aren't looking far enough away.

2. Cross your eyes and get three sets of ovals. Now, in the middle set, the solid center oval should be farthest away. The largest oval should appear closest.

3. Work back and forth, looking through the chart and then crossing your eyes. With practice, each time you look through the chart, the center oval of the middle set should float closest to you; each time your eyes are crossed, the outside oval should float closest.

4. Now it's time to see *bigger, deeper,* and *stronger* at the same time. Normally, we see things at the same distance where our eyes are pointing, and we point our eyes at the distance where we imagine things to be. For instance, if you look at your fingertip, you'll point both eyes at the fingertip. If you straighten your eyes to look at the wall across the room, you'll point your eyes at the wall across

Strong-Sight V

the room. In this step, we are going to make use of this hookup in our brains between where we point our eyes and where we see things to be positioned.

Look through the Strong-Sight V chart until you get three sets of ovals as you did in step 1. Your eyes are now actually pointing at some distance farther away from you than the chart. If the brain hookup we just mentioned is working, you will see the middle set of ovals floating wherever your eyes are pointing. Thus, since your eyes are pointing farther away than the actual page of the book, you should see all three ovals of the middle set floating in space, about two or three feet beyond the actual page of the book.

To appreciate this floating you'll need to use your Big-Sight muscles as well as your Strong-Sight muscles. To help you to see big while you look through the chart, shake the chart slightly and be aware of the whole room out of the corners of your eyes. Compare the position of the ovals to other objects around the room. Try to see the center set of ovals float where your eyes are pointing; i.e., several feet farther away from you than the book. This may be easier to accomplish if you photocopy the Big-Sight V chart and cut it out so that you are able to see around the chart's edges more easily.

5. When you can look through the page and see the center set of ovals floating several feet beyond the book's actual position, now try the same thing with your eyes crossed as in step 2. With your eyes crossed, they will be pointing closer to you than the actual page of the book. Thus, if you are seeing *bigger* as well as *deeper*—and that hookup in your brain is working—you will see the middle set of ovals floating out in the air about six or eight inches in front of the actual page of the book. To see the ovals in this position, hold the book with your right hand and position the index finger of your left hand about six inches closer to you than the page of the book. By comparing the middle set of ovals with your finger, you should be able to see the ovals floating out toward you.

6. When you have mastered looking through the book and seeing the ovals farther away and then crossing your eyes and seeing the ovals closer, there is one more step. Position your eyes so that you see the three sets of ovals, as in the previous two steps. Hold the book with both hands. Extend your arms until they are straight. Move your arms in a circle—either clockwise or counterclockwise. While making this circular movement, keep the bottom edge of the book parallel to the ground. As you rotate the book in this big circle, continue to see the three sets of ovals. Alternate between having your eyes crossed, so that the middle set of ovals comes out toward you, and looking through the chart so that the middle set of ovals floats away from you. When you've mastered this step, you're a Strong-Sight expert.

Strong-Sight I through V are all designed to increase your endurance when using your eyes. In addition, Strong-Sight I, IV, and V should help you to see bigger and deeper.

Strong-Sight I is the least complicated to perform. Strong-Sight V will push you to your limits, but it is an expected skill for those we train in our office. Remember, if you have any difficulty on any step, go to an earlier step or exercise. Don't waste time on a step or exercise that you can't perform.

Even if Strong-Sight I is the only of the five Strong-Sight exercises you do, you will still be able to increase your endurance and depth.

Of all the visual fitness exercises in the book, the Clear-Sight and Strong-Sight exercises are the most physically demanding, and the most rewarding when it comes to increasing your visual endurance.

As we will discuss in chapter 11, visual fitness trainers have a number of tools at their disposal for training the Strong-Sight muscles. If the exercises in this chapter are too difficult or are not allowing you to reach your visual endurance goals, then by all means visit a doctor who trains visual fitness. It's entirely possible that you have an undiagnosed eye-muscle problem that may require professional assistance to improve.

7

VISUAL FITNESS SKILL V: QUICK-SIGHT

How Fast Can You See?

Think what it would be like if you had twice the time in your day—twice the time for reading, twice the time for hobbies, twice the time to return the serves of your opponent during tennis? What if you could accomplish all this without giving up any more sleep?

Sound good? Consider a story written by Paul Fountain for an issue of *Flying* published in August 1945.

In World War II, soldiers' eyes were falling behind the technology of war. In August 1942, for instance, Canadian, British, American, and Free French commando forces failed in their raid of an eleven-mile stretch of French coast near Dieppe, a town on the English Channel midway between London and Paris. During that raid, ninety-two planes were lost, sixty of them shot down by British antiaircraft personnel who couldn't see fast enough to identify their own planes.

This was no isolated incident of soldiers' eyes failing to keep up with the split-second demands of aerial warfare. There were hundreds of similar accounts. The Italians, for instance, shot down their own air ace, General Balbo.

There was, however, an answer to this problem in the making. Throughout the 1930s, Dr. Samuel Renshaw, a professor of experimental psychology at Ohio State University, had been studying and training the speed of human perception. Using a tachistoscope, an instrument that flashes images at a fraction of a second, he showed in his experiments that it was possible to train students to identify a nine-digit number flashed at 1/100 of a second. One of Renshaw's subjects increased his reading rate from 350 to more than 1,000 words per minute.

With the advent of the war, Renshaw began modifying his instruments to flash the silhouettes of aircraft. By 1942, the United States Navy was looking into Renshaw's visual fitness methods, and by the end of the war thousands of officers, in not only the U.S. fleet but the British, Canadian, and Australian armed forces as well, had been trained. For many, their accuracy at identifying aircraft silhouettes had increased from 20 percent when flashed at 1/25 of a second to 98 percent when flashed at 1/100 of a second.

Not only could the trainees see more quickly, they could see bigger—Renshaw's methods trained their side vision as well.

Frequently, those who had received the training stated that more than their vision had been affected; they felt more intelligent as well.

After the war, Renshaw began working with many of the leading eye doctors who trained what in this book we've called "visual fitness." As a result, his legacy has become a part of the heritage of visual fitness trainers everywhere.

So how about you? How fast can you see? Imagine being able to see faster for reading, driving, sports, and work. How would that improve your life?

In the Big-Sight chapter, we talked about how seeing details allows us to see more; seeing the expressions on the actors' faces, for instance, allows a greater appreciation of performances. The ability to see quickly similarly expands our worlds.

If we're only able to pick up information in half-second chunks of time, and something occurs in a quarter of a second, we won't be able to see it. If, however, we can pick up information in a tenth of a second, then we have more than enough time to see something that flashes by in a quarter of a second. The faster we see, the more we see. Seeing quicker can also increase accuracy and certainty. The faster we can see, the more extra time we have to look. If we can pick up information at a hundredth of a second, then when we're driving and have half a second to see a street sign, we can recheck our perception of that sign many times before it vanishes from view.

There are many examples of the advantages of seeing quickly. The professional baseball player who can hit a fastball approaching at one hundred miles per hour has to see fast. During reading and computer work, the faster we can see, the faster we can finish, or the more we can accomplish. In driving, the quicker our sight, the safer the trip. On a two-lane road, for instance, a car coming over the dividing line can be hurtling closer a whole lot faster than the fastest fastball that a professional ballplayer will ever encounter. When forced to react in such a situation, there's no time for seeing slowly.

Seeing quickly is a learned skill. Seeing quickly can be trained. You may not be interested in seeing enemy planes, but if you're ready to see more quickly for reading, sports, or driving then you'll want to add the following Quick-Sight exercises into your visual fitness workouts. If you're in more of a hurry to see faster, you can spend as much time as you like on Quick-Sight because, unlike Clear-Sight and Strong-Sight, there is nothing about Quick-Sight that will tire your eyes out more than any other type of seeing.

If you're reading this book through for the first time, and you'd like a quick introduction to Quick-Sight, then skip ahead and take a quick peek at the arrows in Quick-Sight II.

Your Quick-Sight Workout

A tachistoscope is an instrument for testing or training speed of seeing. The word tachistoscope was coined in the late nineteenth century and was derived from Greek words meaning "swiftest" and "target." The original instruments used shutters like a camera to flash targets for a fraction of a second. More recently, there are computer programs that accomplish the same thing. For the purposes of this book, however, you are going to make your own simple tachistoscope flasher from a piece of paper.

Fold a piece of standard paper (8½" × 11") in half so you have a double-thickness piece of paper approximately five and a half inches tall and eight and a half inches wide. Use a knife or scissors to cut out a rectangular window through the two layers in the middle of the sheet of paper. Make the window approximately three-eighths inches tall and two and a half inches wide as shown below. If the window is correct in size, you should be able to place the opening over the numbers below and completely see all seven numbers.

3 9 2 7 5 1 6

With your flasher constructed, you're ready to begin your Quick-Sight workout.

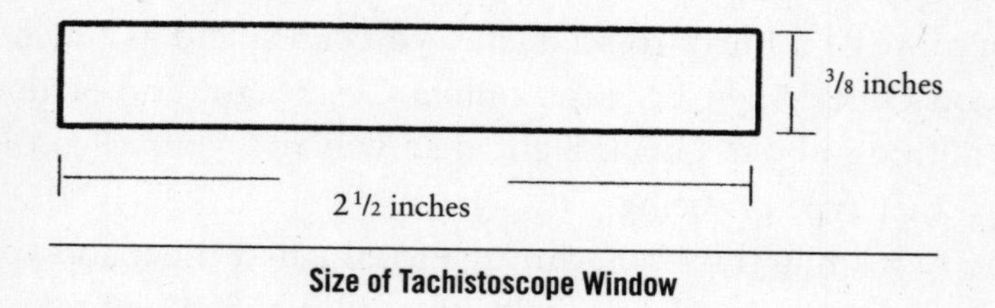

Size of Tachistoscope Window

Quick-Sight I

On pages 108 and 109, you will find two columns of three-digit numbers. In the first column on page 108 the beginning number is 527. Hold the window above the 527 until you can see the hyphenated word *Three-Digit* in the right-hand side of the window. Move your piece of paper slowly downward until the window has passed over the 527 and rests in the blank space between 527 and 394. By moving the paper slowly, you should have easily seen the 527 as the window passed over it.

Place the book flat on a table. If you are right-handed, hold your homemade flasher with your right hand and steady the book with your left hand. If you are left-handed, hold the flasher with your left hand and steady the book with your right.

Return the window to its position above the number 527. Practice moving the window down across the number at increasing speeds until you can just barely see the number "flash" as the window passes over it. The trick you'll need to learn is to move the flasher quickly but halt it in the space between one number and the next so as not to confuse the two numbers.

Once you've learned to use your "flasher," begin to work down the first column of three-digit numbers. Use the following four steps:

1. *Preset*: Relax. Open up your side vision to include the room on both sides of your book. Point your eyes to where the middle of the window will pass over the numbers. You want your eyes pointed at the middle of the numbers, not at the first number, because otherwise you may not be able to see the far-right digits—especially as the numbers grow longer in later exercises.

2. *Flash*: Flash the number, by moving the window downward over it.

3. *Process*: While the number is still covered by your paper, the window resting just below the number, try to see all the digits in your mind and call out the number to yourself. If you're not sure, repeat step 2 until you are. (Note: The ability to see pictures in your mind is covered in detail in chapter 8, the Smart-Sight chapter. If you can't yet see the digits in your mind, just say them to yourself for the time being and continue with the Quick-Sight exercises anyway. As you keep working, be aware of any digits that you do see in your mind. It's likely that as you go on you will be able to see more digits—especially if you relax. You will learn more in the Smart-Sight chapter that will also help with your Quick-Sight.)

4. *Compare*: Check your accuracy by moving the flasher higher on the page until its window frames the number. Compare what's in the window to the picture in your mind—or the numbers you are saying to yourself should you not have a picture in your mind. If necessary, you may find it easier to write the digits down on a sheet of paper after you have flashed them. Then, when you raise the window, you can compare what you've written to what you now see.

Complete the Preset, Flash, Process, and Compare steps with each number in both columns. Work with the numbers on page 109 in the same way. Repeat all the three-digit numbers while trying to increase the speed of your flashes until you can recall correctly each number when you've flashed it for the barest fraction of a second. If you find yourself memorizing the numbers, work upward from the bottom of the pages.

When you have mastered the three-digit numbers using very quick flashes, move to the four-digit numbers on pages 110 and 111. Work these using the same four steps. Move through the four-digit numbers as many times as it takes until you can perceive them at the fastest of flasher speeds without making more than one mistake on a page.

Perform the five-, six-, and seven-digit numbers in the same way (see pages 112–119). If you're having difficulty getting the last digits, remember to keep your attention on the middle digit not the first. Also, be sure you're performing the Preset step, by relaxing and opening up your side vision.

Each time you start a page where the numbers are longer, slow down your flasher to allow yourself more time. If after you have worked through the longer numbers a time or two and you've tried slowing down and you still can't identify the numbers, then return to the previous set of numbers. Continue to work that set at faster and faster speeds until you are ready to return to the longer numbers.

When you are doing the longer numbers, you may recall them more easily if you break them down into two shorter numbers in your mind.

For instance, instead of trying to recall 315327 try to recall 315 followed by 327.

The farther away you hold the book, the less you have to open up your side vision to see all of the digits. For this reason, you may want to begin by holding the book farther away. As your skill improves you can open up your side vision more and move the book closer.

Quick-Sight II

In Quick-Sight I, we worked on learning to see faster and asked ourselves the question, What is that? To be more precise, we were asking ourselves the question, What is that number? In Quick-Sight II, we are going to continue to work on speed, but we're going to ask the question, *Where* is that? Specifically, Where is that arrow pointing?

First make a new flasher by taking a piece of standard paper (8½" × 11") and folding it in half as you did for the first flasher. But don't cut a window in this one.

Turn to the charts on pages 124–126, which contain rows made up of three arrows each. Align the top edge of your new flasher with the top boundary line above the arrows. Move the flasher down quickly to expose the top row of three arrows, then quickly upward to cover it again: two quick movements—one down then, immediately, one up.

Practice this flasher movement a number of times on the top three arrows until you can make the movement easily and quickly so the arrows are exposed for only an instant. When you've mastered your new arrow-flashing technique, you're ready to begin the same four-step procedure you used in Quick-Sight I:

1. *Preset*: Position the flasher's top edge above the set of arrows with which you want to work. (Since you just practiced with the first row and know which way those arrows point, place the flasher's top edge below that row.) Relax. While you point your eyes where you imagine the center of the next row will be, open up your side vision; be aware of the room, not just the page of the book.

Three-Digit Numbers

5 2 7	4 6 2
3 9 4	2 7 1
9 2 6	4 3 2
1 8 4	8 6 9
8 2 1	5 1 7
4 9 2	7 3 8
6 7 3	1 3 6

4 8 2	3 9 1
8 3 4	7 8 6
1 3 1	5 1 9
2 5 5	4 7 2
3 4 5	2 2 5
7 9 3	1 5 4
5 7 8	7 3 1
8 2 1	8 1 3

Four-Digit Numbers

3 8 5 2 5 4 2 1

9 7 4 1 1 3 2 5

8 6 3 2 3 7 4 7

7 1 8 9 2 9 8 4

2 3 2 1 7 3 9 5

1 9 2 8 4 5 5 3

4 1 8 7 9 3 2 4

5 1 3 9 9 2 8 3

4 1 9 2 1 7 5 9

3 2 2 8 2 4 1 9

7 7 9 8 3 1 1 4

9 7 9 9 1 5 5 5

4 5 1 3 4 3 8 7

1 8 5 1 5 8 2 1

2 3 3 4 7 2 3 2

Five-Digit Numbers

4 4 4 5 4	7 1 2 9 8
1 2 1 1 1	9 7 4 1 3
7 7 3 7 3	4 8 7 2 1
5 1 5 1 5	3 9 4 8 5
7 7 2 2 1	5 2 3 7 7
1 1 5 5 5	8 3 1 3 4
2 3 4 4 1	2 5 5 5 9

5 7 4 8 5   3 9 1 8 4

4 1 9 2 2   2 3 5 4 7

8 2 8 2 1   5 9 8 5 1

3 8 2 3 7   7 8 4 3 9

2 3 5 5 8   1 7 9 9 2

1 1 7 1 1   4 4 7 7 5

7 8 4 1 3   9 7 2 3 1

Six-Digit Numbers

1 1 2 1 1 1

5 5 5 2 2 2

4 3 4 3 4 3

7 7 7 7 2 7

9 2 2 9 2 2

1 2 3 1 2 3

5 9 9 5 9 9

3 1 5 3 2 7

4 1 9 8 3 9

7 7 1 1 7 7

4 1 8 4 1 5

2 5 3 7 4 9

5 1 9 8 3 5

1 1 8 4 1 1

3 8 9 1 1 4

7 5 1 9 8 2

1 9 2 5 5 7

9 9 8 1 3 2

2 1 7 7 8 4

4 1 7 5 5 1

5 1 4 8 2 3

4 9 9 1 5 5

3 7 2 2 7 7

Seven-Digit Numbers

1 1 1 5 1 1 1

9 9 2 2 2 9 9

5 2 2 1 2 2 5

4 4 3 3 3 4 4

2 2 2 9 3 3 3

7 7 1 2 1 7 7

1 1 1 1 1 3 1

8 8 6 2 2 1 1

9 9 2 2 7 7 8

5 5 5 3 8 2 1

3 3 2 3 3 2 7

9 7 2 2 2 7 9

3 9 1 9 4 3 3

4 4 7 2 9 8 1

1 8 1 8 2 2 2

8 1 2 9 3 5 7

7 1 5 2 8 8 4

3 9 1 5 3 4 1

1 3 2 8 8 4 1

3 8 5 7 4 1 9

9 5 2 1 7 8 4

3 1 5 4 9 8 7

1 9 5 7 3 8 2

2. *Flash*: Flash the next row—a quick downward motion followed by a quick upward motion. Again, try to keep your eyes pointed in the center of the row of arrows rather than using your eyes to follow the flasher.

3. *Process*: In your mind, try to picture which way the arrows pointed. You can say the directions of the arrows to yourself, but also try to see them in your mind. After you've seen them in your mind, you can draw them.

4. *Compare*: If you're not sure you saw them correctly, repeat steps 1 and 2 until you think you've got it. Then move the flasher down and leave it down to check your answer.

Work through all the three-arrow targets, then turn the book upside down and repeat the process. In this manner, you'll double the number of patterns available for practice. If you found these patterns too easy on your first attempt, try to increase the speed of your flash on the second.

When you can easily do the three-arrow targets while flashing them as quickly as you can, move to the four-arrow targets. If these are too difficult, return to the three-arrow targets. Work these until they are more second nature. Then, move on to the four-arrow, five-arrow, and six-arrow targets on pages 127–135 in the same way.

As when you were working with the digits, the main errors you will make are 1) not looking in the center of the target and 2) being too central in your seeing rather than being aware of how much room is visible around the edge of the book. This opening up of your peripheral vision will become especially important when you come to Quick-Sight III.

Optional Steps: For Those Who Get Right and Left Confused

If you have always been confused when trying to tell left from right, and you have to pause and think when asked to turn right or left, then now is the time to become unconfused by using a procedure based on the work of Dr. Robert Pepper from Lake Oswego, Oregon.

1. Before working with the charts on pages 124 to 135, practice going through the chart on page 127 in the following manner: lay the book flat on a table and touch each arrow with the tip of your left index finger while moving your right hand in the direction of the arrow as if that hand were conducting a symphony. Simultaneously, call out that arrow's direction (up, down, right, or left). Work through each line in this manner, keeping your place with your left hand, mimicking the direction of the arrows with broad sweeps of your right. Repeat this process a number of times, each time increasing your speed until you can move your right hand and rattle off "Up," "Down," "Right," and "Left" without the slightest hesitation. Try to establish a rhythm as you work and maintain that rhythm throughout the entire figure.

2. Once you've mastered step 1, work through the figure again but this time when you touch an arrow with your fingertip and move your right hand in the direction of the arrow, call out the opposite direction. For instance, if the arrow is pointing upward, move your right hand upward but call out "Down." If the arrow points left, move your hand left but call out "Right." If right and left are not yet ingrained in your thinking, you will fall apart on this step and may have to return to step 1. Persevere. As you learn to make these errors on purpose, right and left will become a part of you.

3. Finally, return to calling out the correct direction, but move your right hand in the opposite direction. For instance, if the arrow points upward, call out "Up," but move your hand down. If the arrow points to the left, call out "Left," but move your hand to the right.

It may take some time to master these steps, and steam may come out of your ears as the mud puddles in your brain evaporate, but it's easier than spending the rest of your life confused by right and left. By the time you can accomplish steps 1–3 with ease, you'll be able to tell your right hand from your left hand without having to think about it. You'll also put a serious dent in your lysdexia (pardon me, your dyslexia). Most importantly for our purposes, however, you'll be ready to succeed with Quick-Sight II and III.

Quick-Sight III

In this section, the emphasis is on seeing even bigger at the same time you are seeing faster. Perform this exercise exactly as you did Quick-Sight II, but instead use the charts on pages 136–155.

Work through the five figures with three arrows, then rotate the book 90 degrees so that the book is sideways and work through them again. Rotate the book another 90 degrees so that it is now upside down. Repeat the five patterns before rotating the book another 90 degrees. By the time you have rotated the book into all four orientations, you will have worked with twenty patterns of three arrows rather than five patterns.

Each time you flash a page, try to see a picture of the arrows in your mind even if you have to say their direction to yourself to support the picture you're seeing. When you are solid on the three-arrow patterns, repeat the entire rotation process with the four-arrow patterns, the five-arrow patterns, and the six-arrow patterns found on pages 127–135.

As with Quick-Sight II, initially you may have to flash a pattern several times. Just be sure that you keep your peripheral vision open so that you can see as much of the room as possible out of the corners of your eyes. Also, be sure you keep your eyes pointed to the center of each figure so that you have the best chance of seeing all four corners at the same time.

If you have plenty of paper and a pencil available, a better way to work Quick-Sight III is to fold a number of pieces of standard-size paper in half to make flashers. Use one of these flashers at a time. Flash a chart then lay the flasher on top of the pattern and draw the four, five, or six arrows in their correct positions and directions, so that if you had x-ray eyes your drawn arrows would be lying exactly upon their respective hidden arrows. This accomplished, move your drawing to the side and compare it to the figure to see how you did. You may have to flash a target several times to get the picture of the arrows in your mind, but once you begin to draw, don't lift the flasher until you are finished with your drawing.

Use all three exercises—Quick-Sight I, II, and III—in your workouts. Don't try to master Quick-Sight I before you start Quick-Sight II and III. Instead alternate among the three: exercise with Quick-Sight I during the first workout, Quick-Sight II the next, and Quick-Sight III on the third workout. Keep progressing on I, II, and III as long as you are successful. If you run into a snag, back up.

Mastering Quick-Sight I, II, and III may take some time, especially if you're spending only seven minutes on your Quick-Sight workouts. As mentioned earlier, you can spend as much time as you wish on these speed-seeing drills without unduly fatiguing your eyes. Do not, however, abandon your other visual fitness muscles to use all your exercise time on Quick-Sight. This would be a little like abandoning all your bodybuilding exercises except sit-ups. The exercises work together to allow you to take in more information in less time with less effort. The Big-Sight and Deep-Sight exercises, for instance, help with relaxation, and this relaxation is key in developing your ability to picture images in your mind. I designed the Smart-Sight exercises in the next chapter to help if you're having difficulty with Quick-Sight I, II, and III.

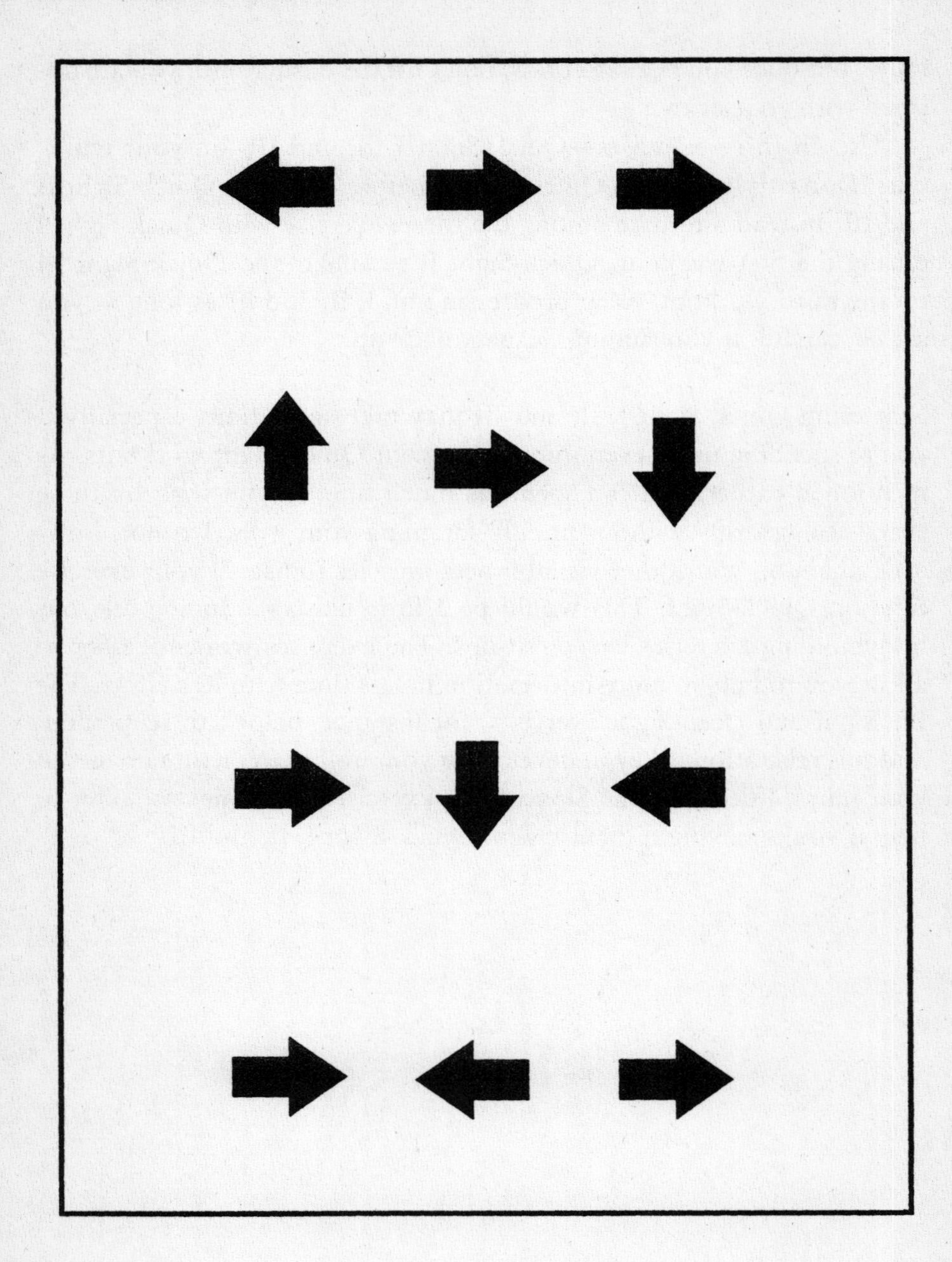

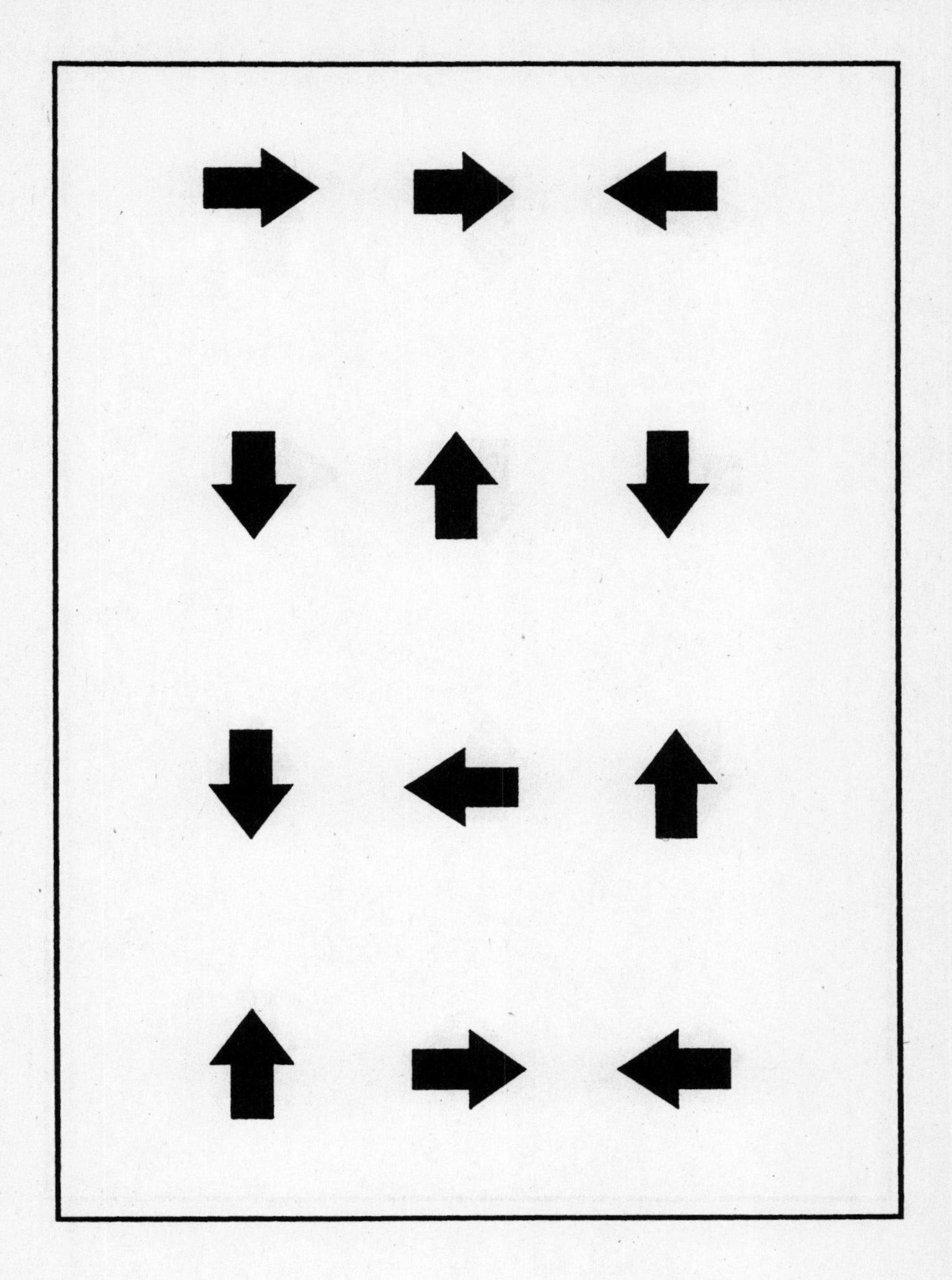

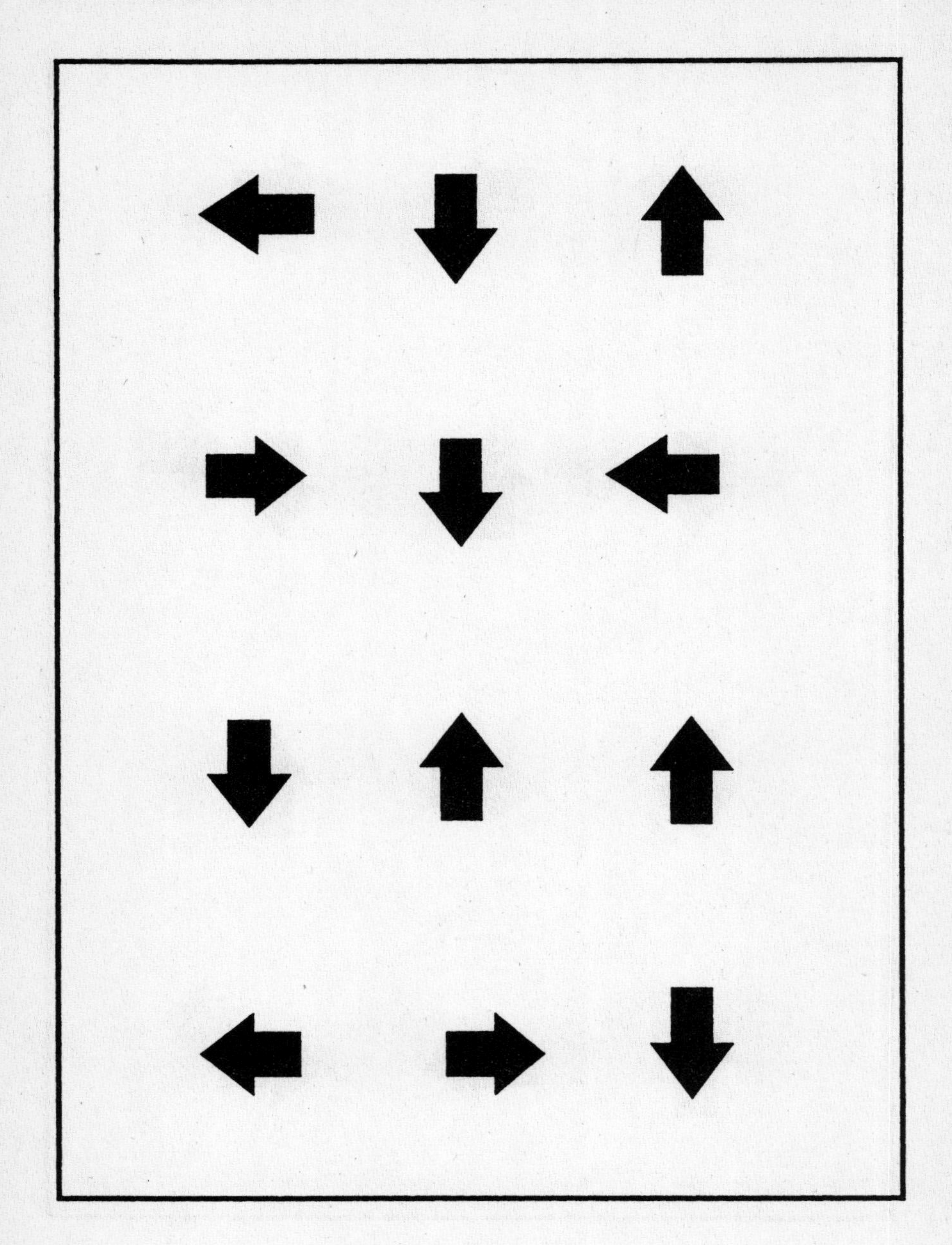

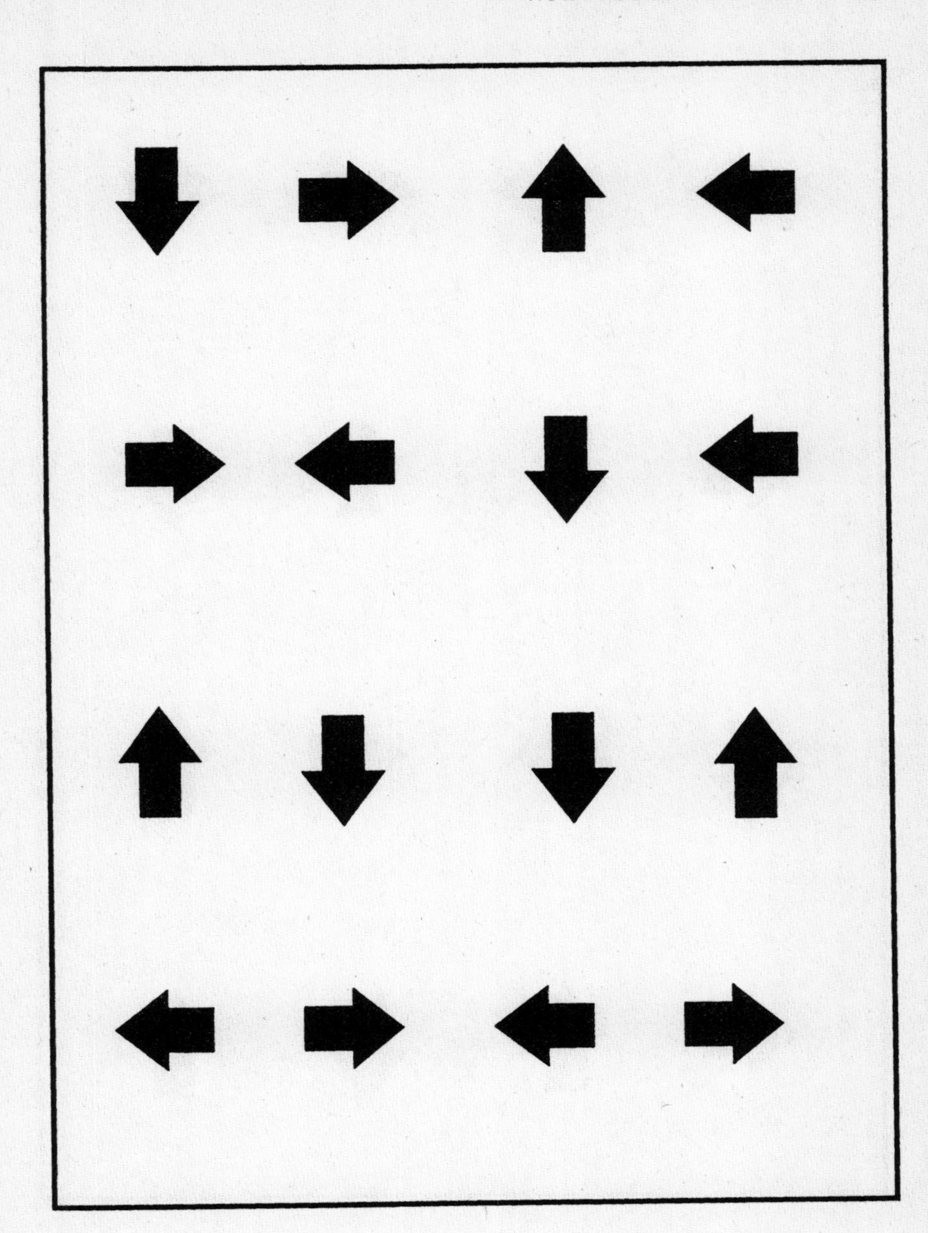

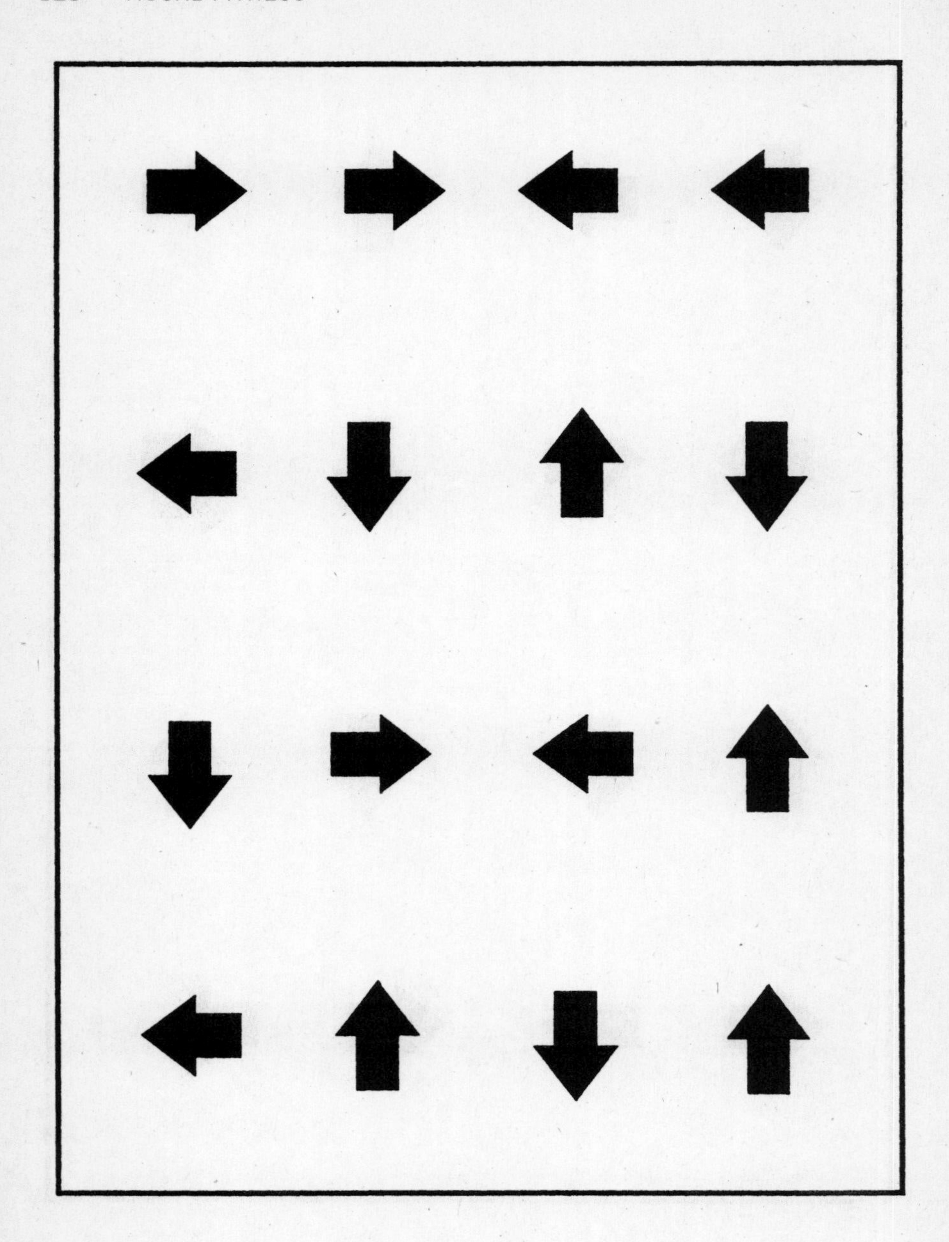

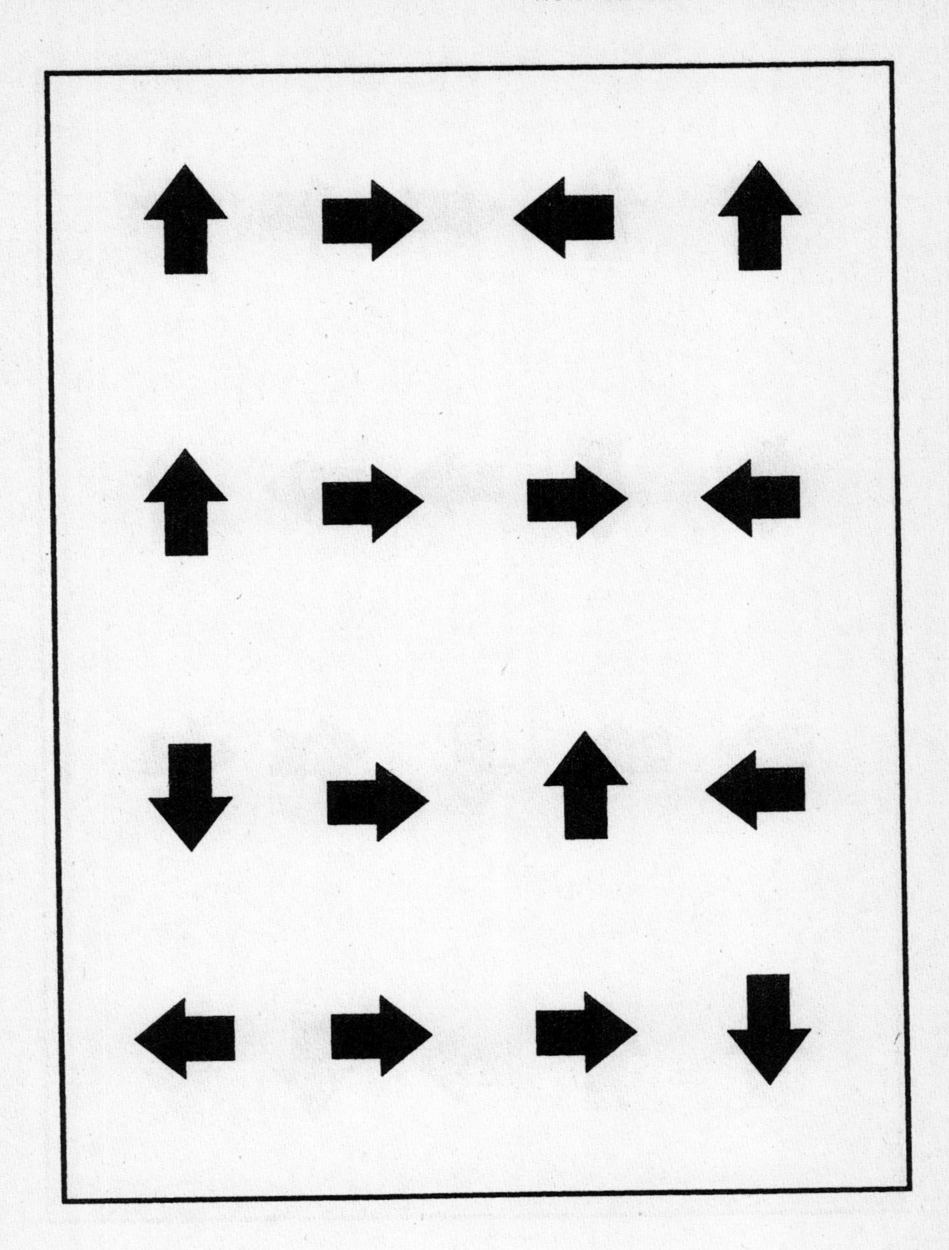

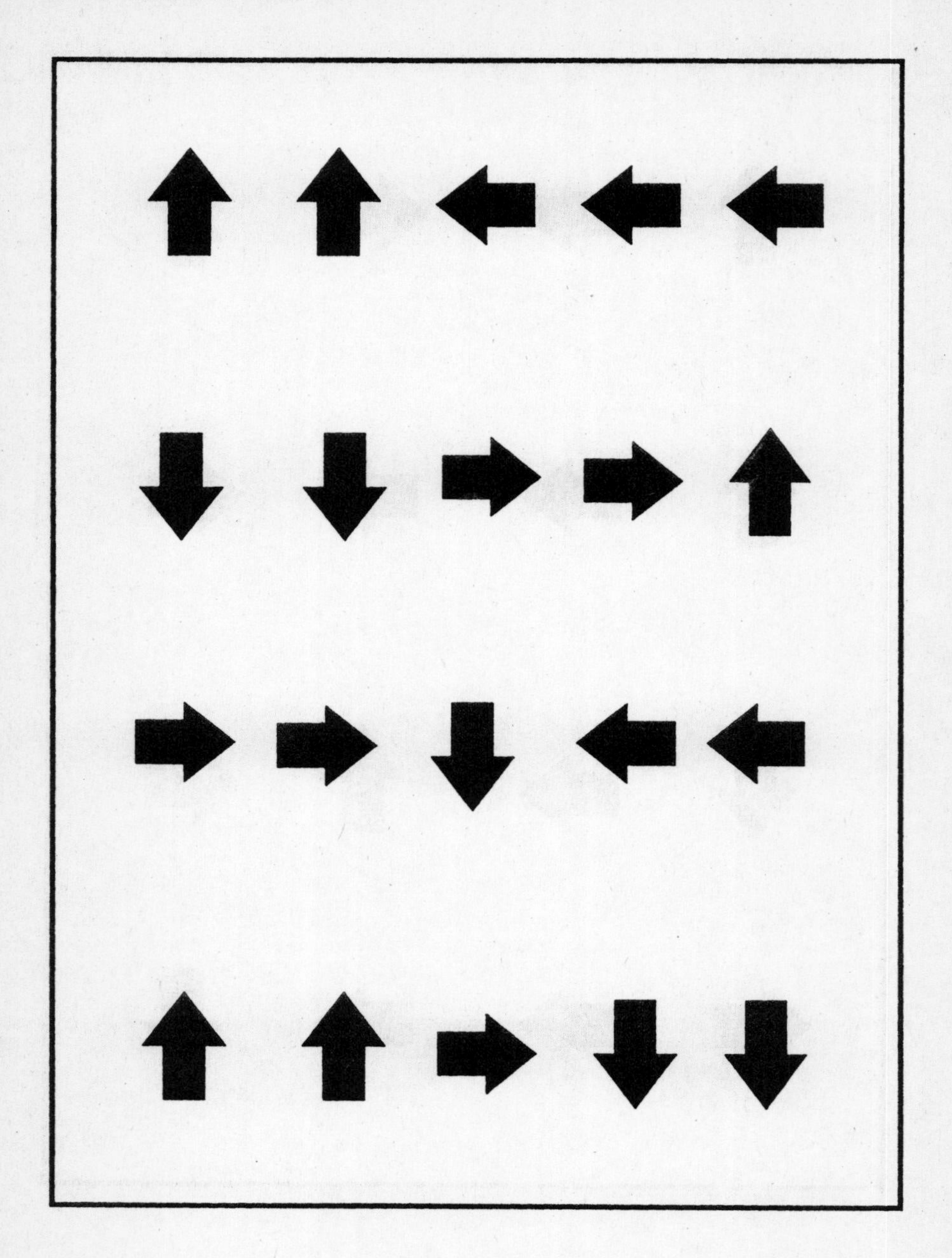

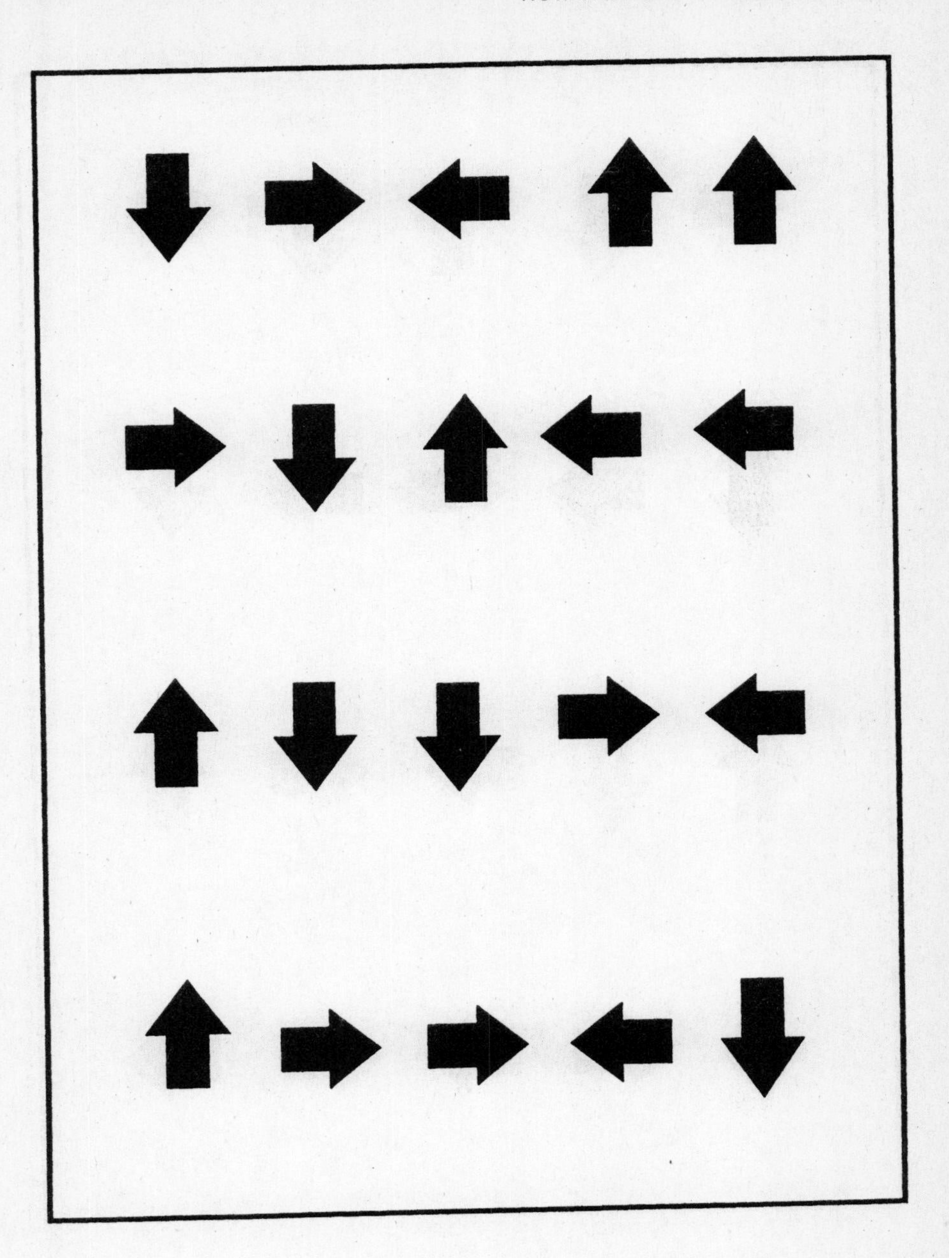

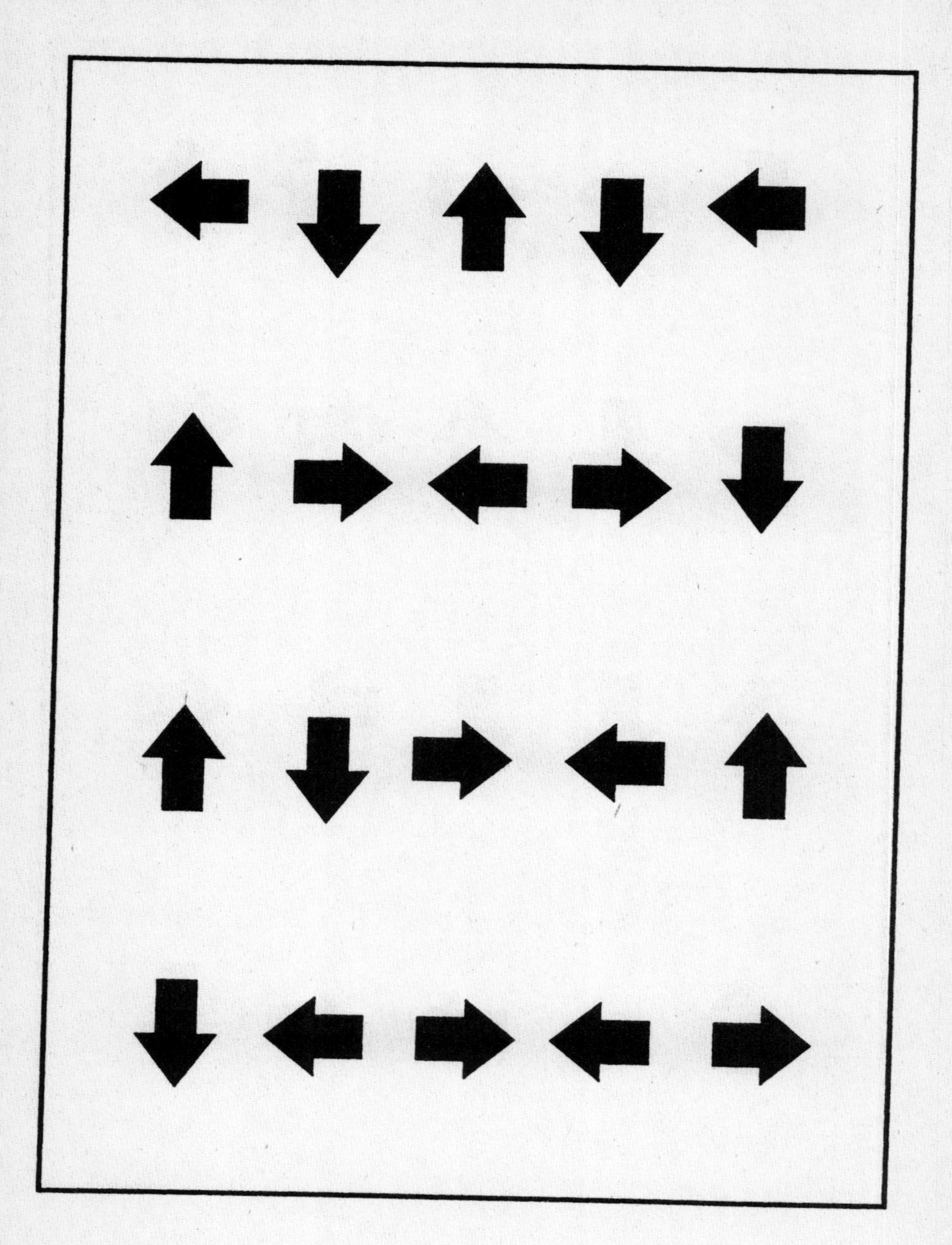

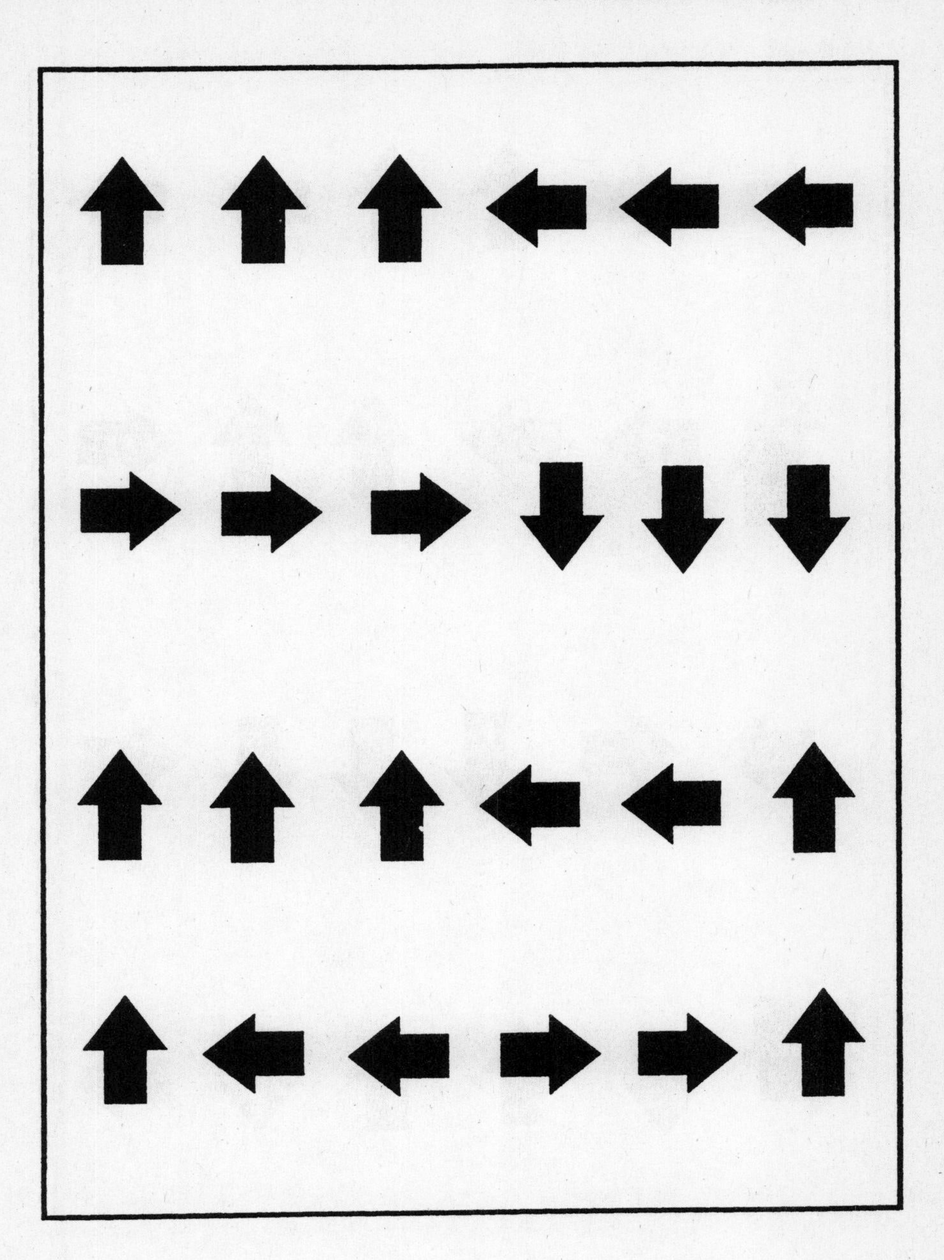

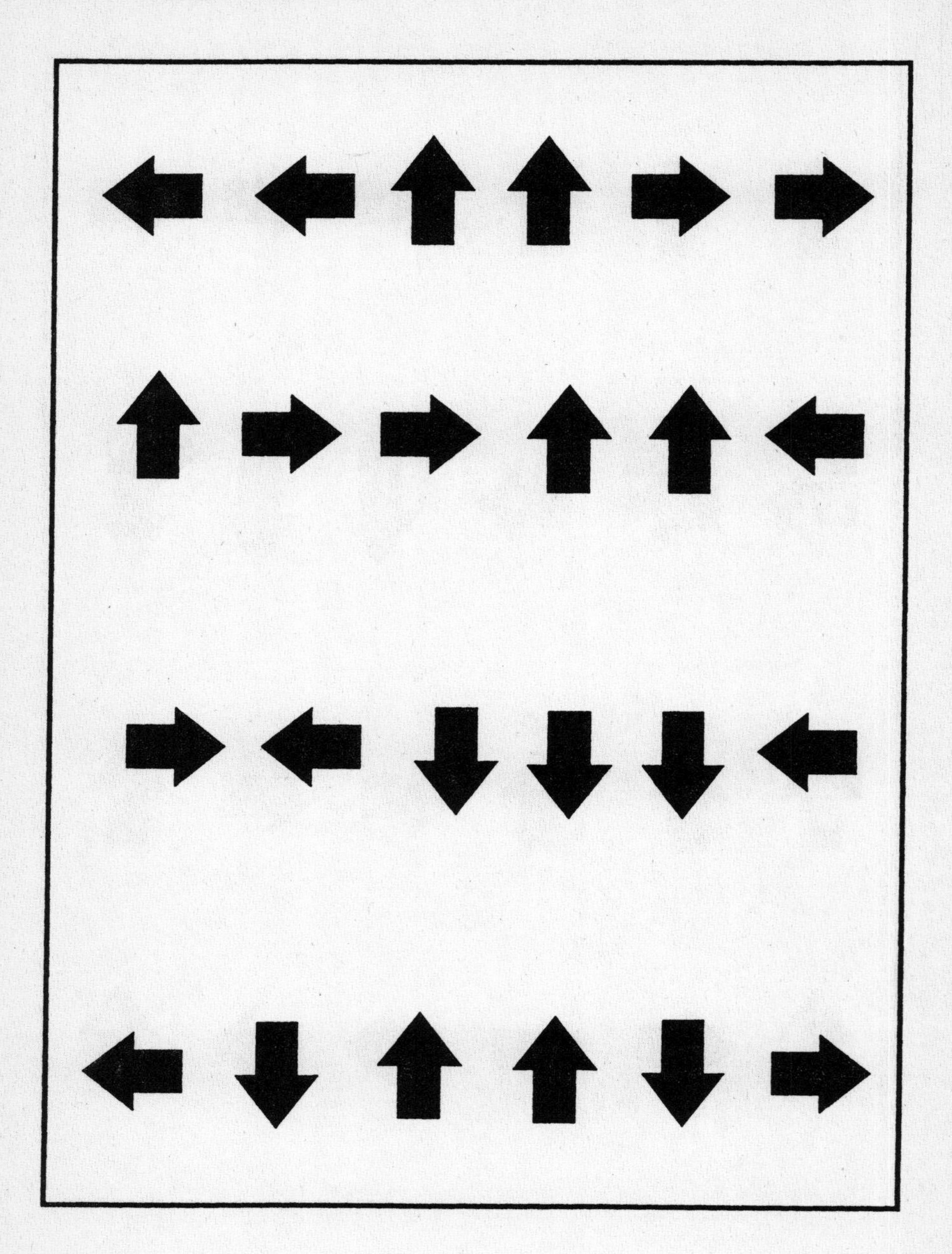

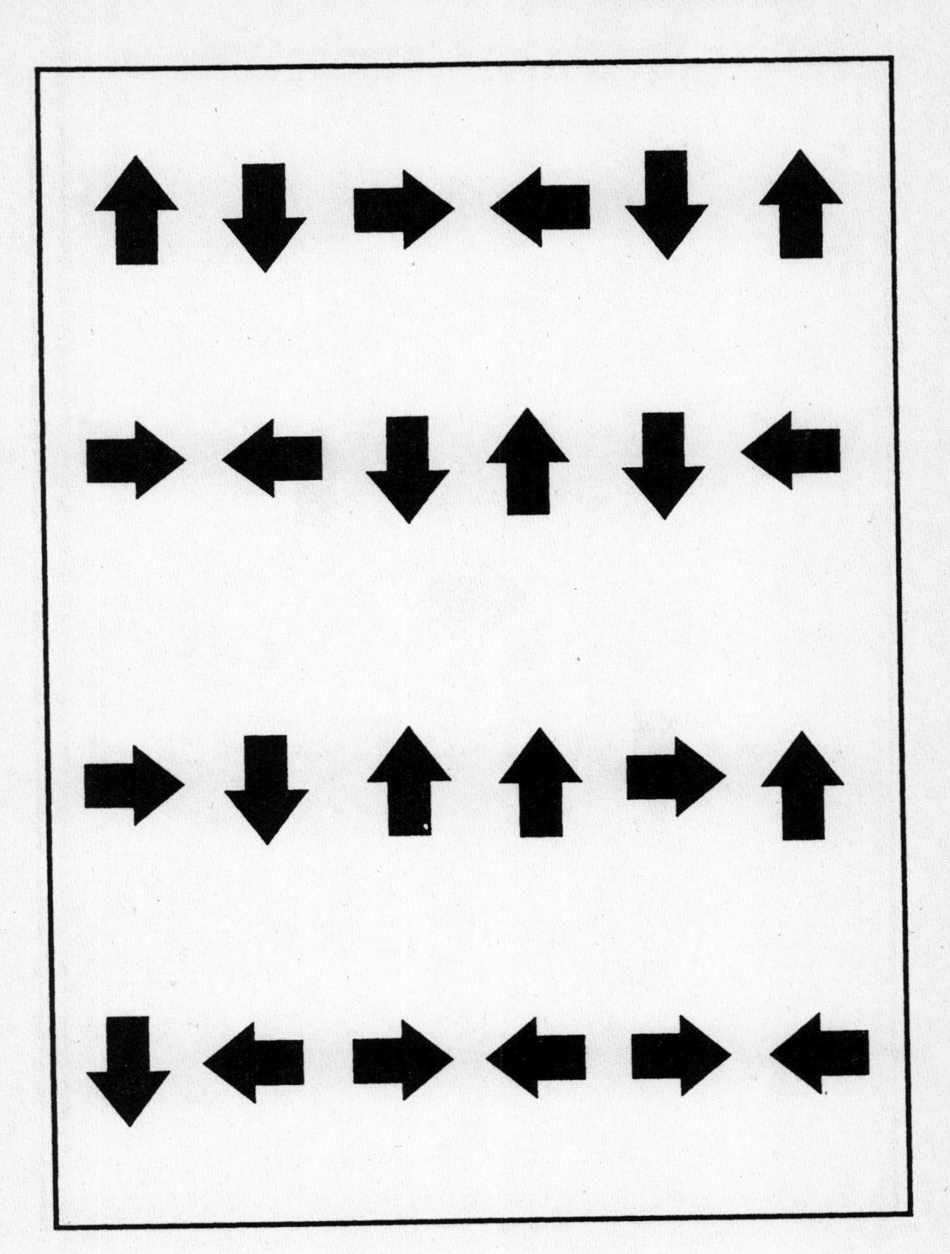

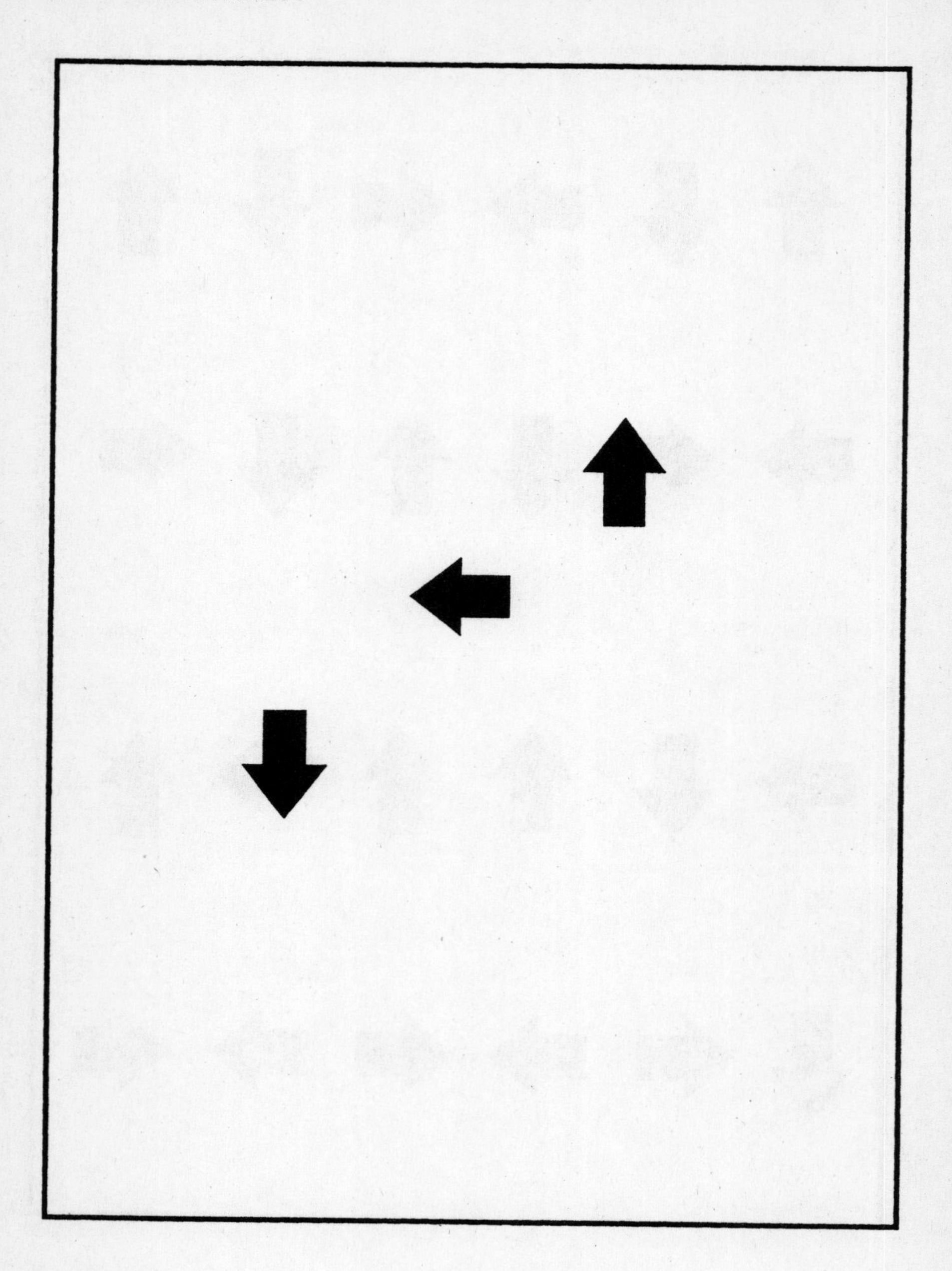

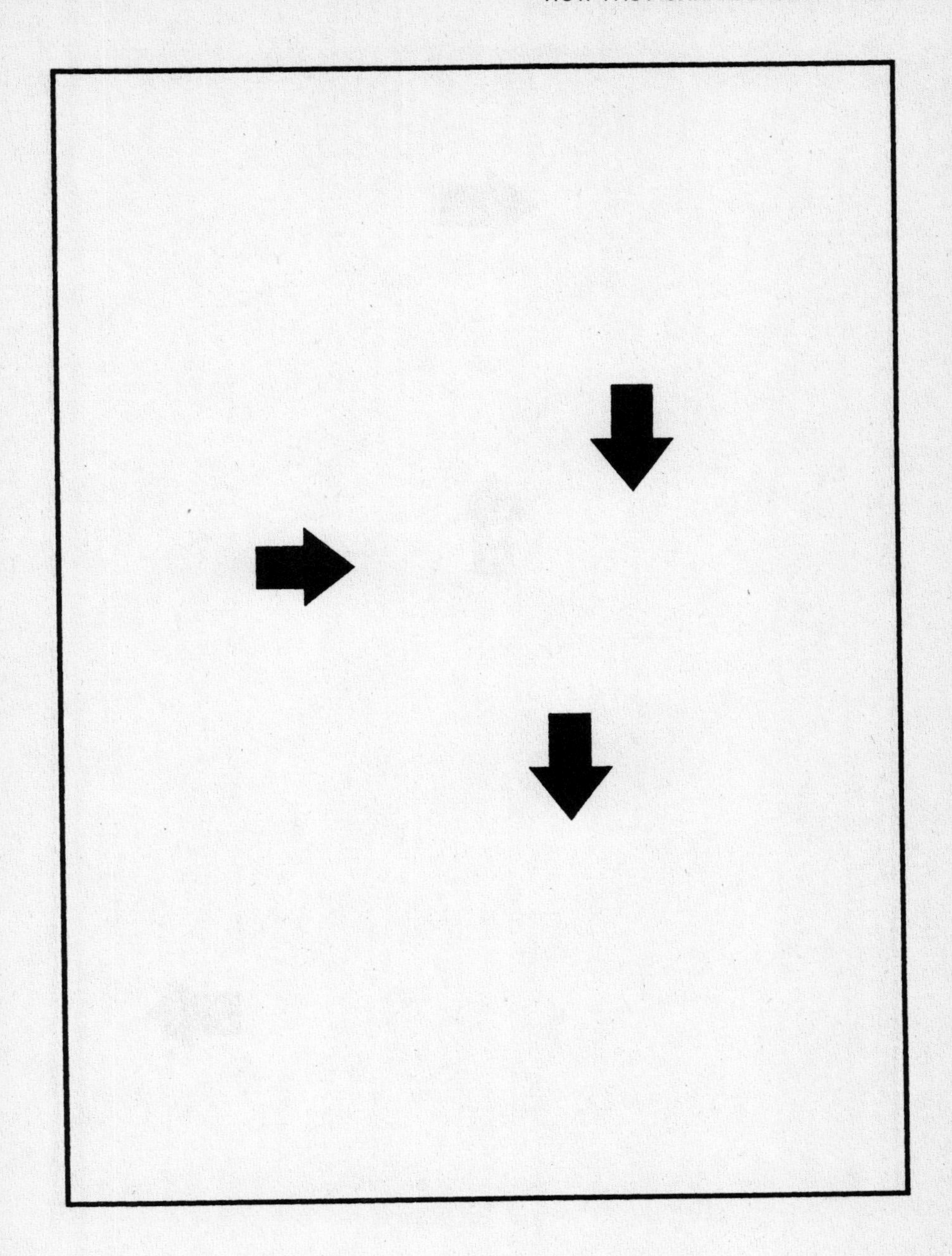

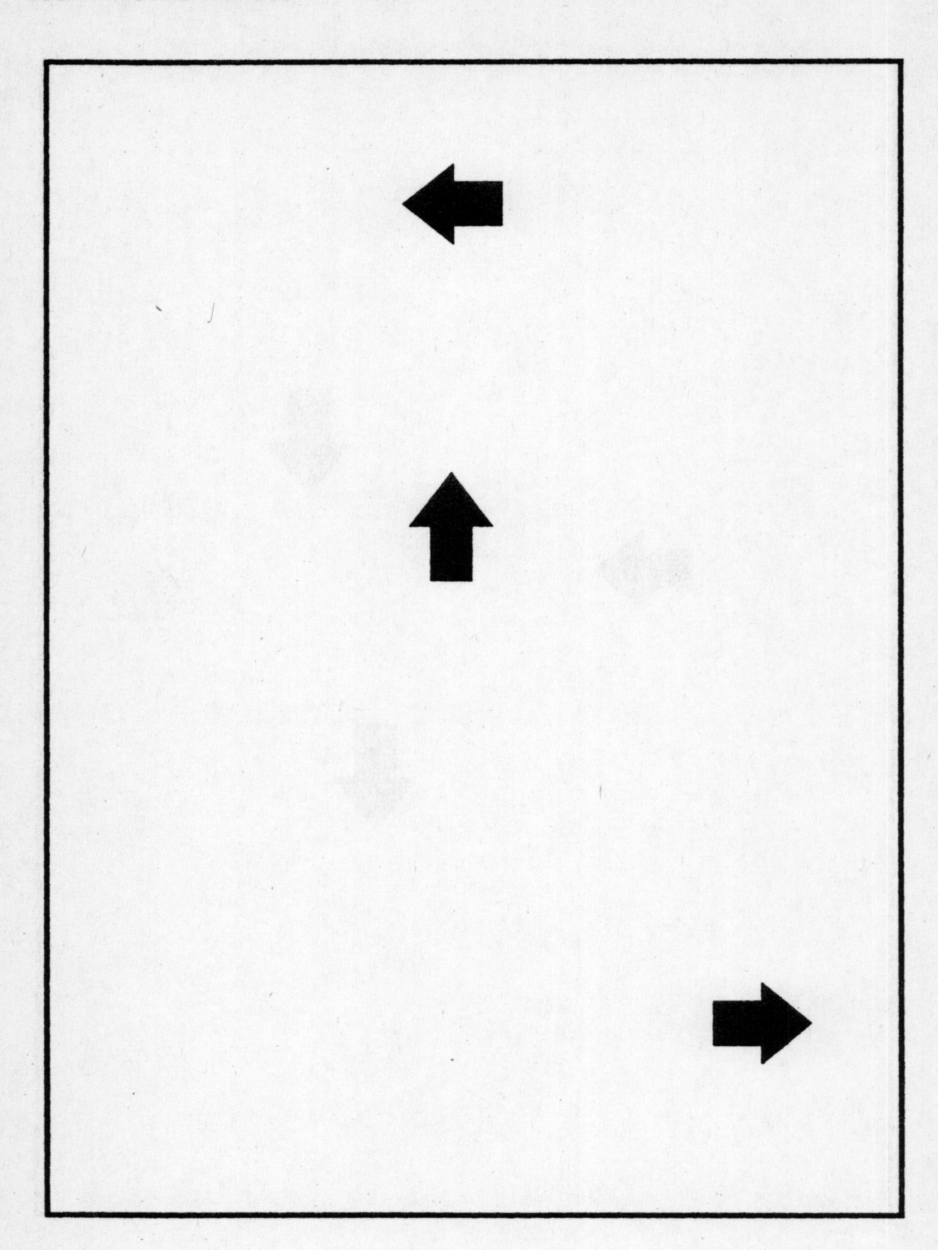

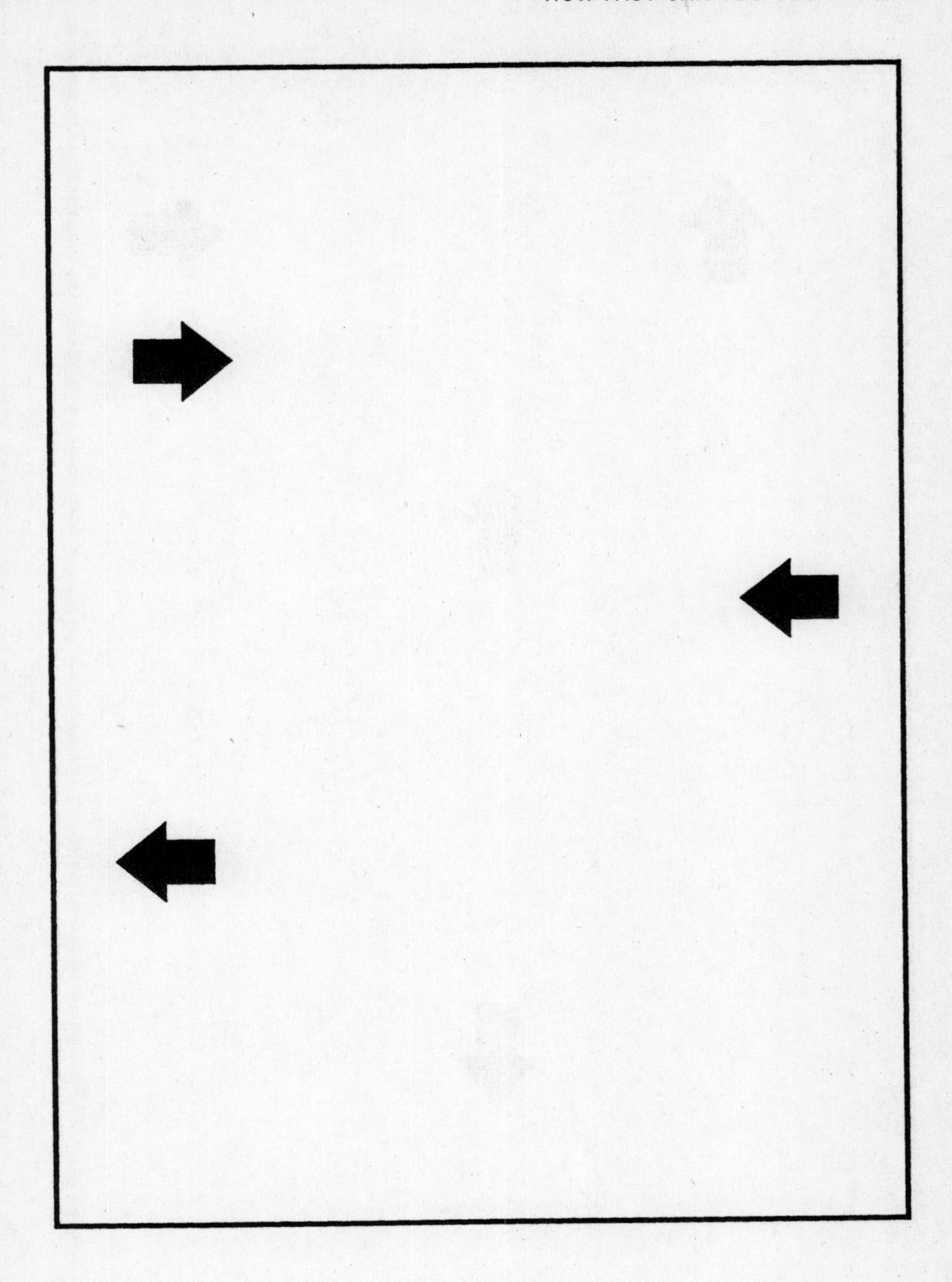

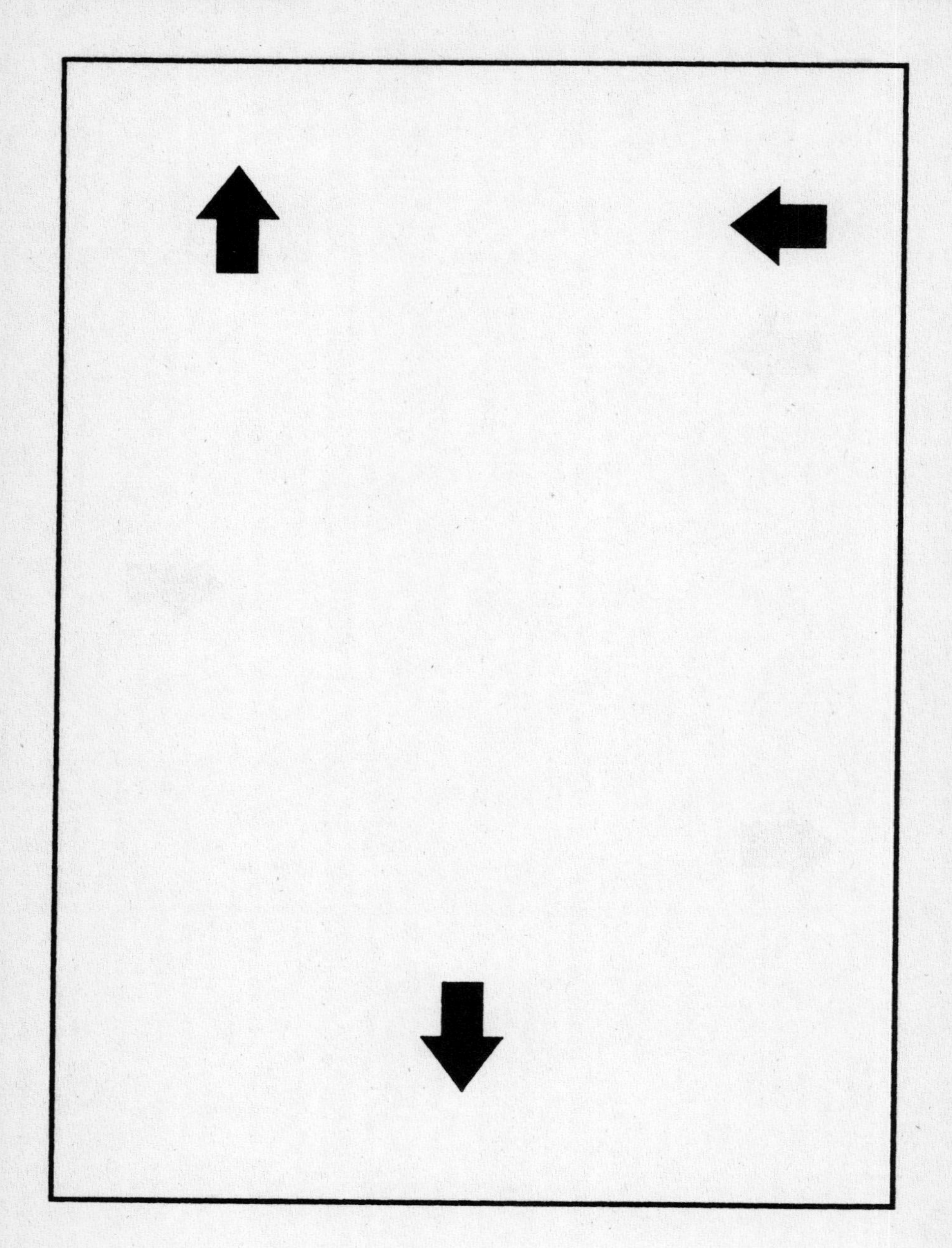

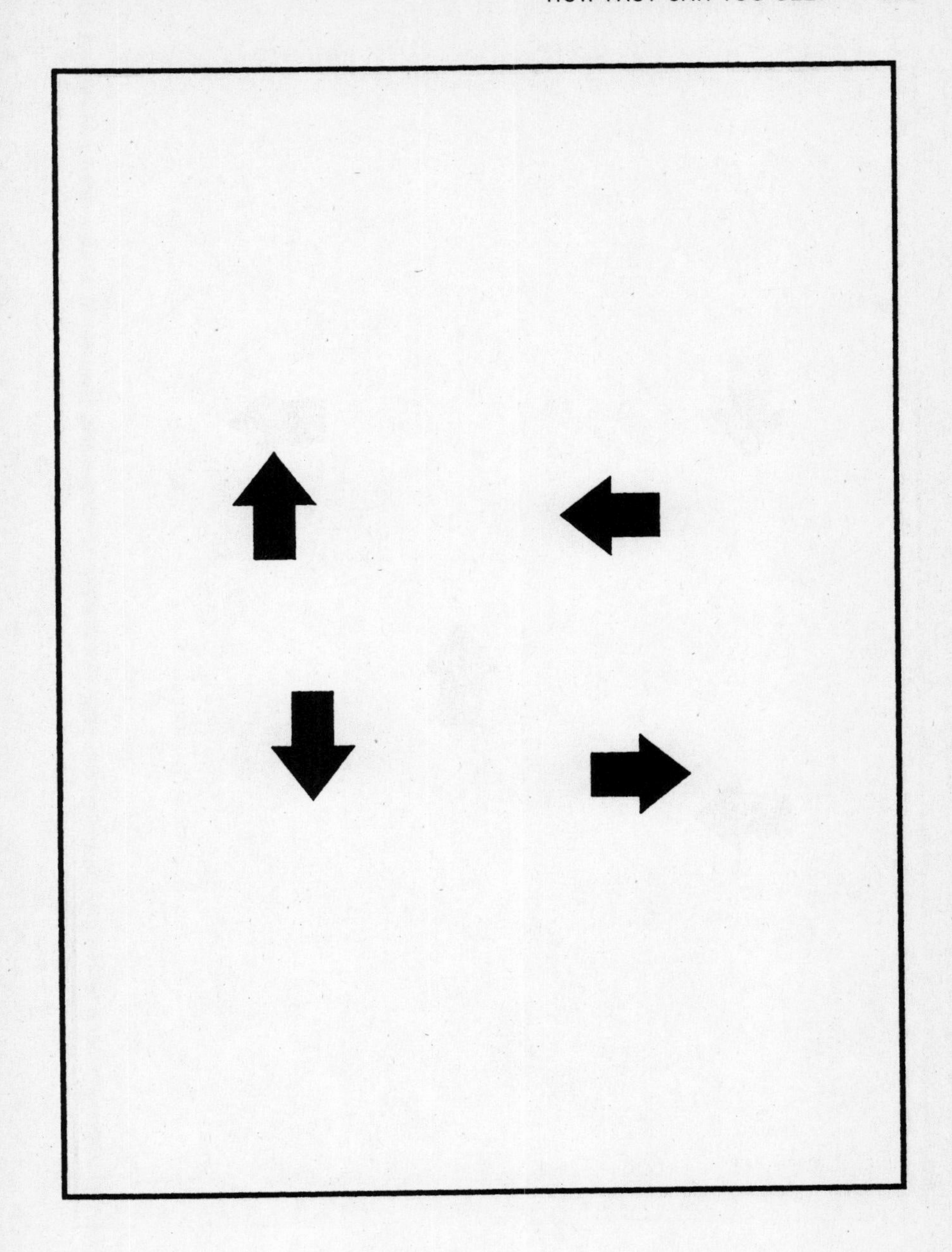

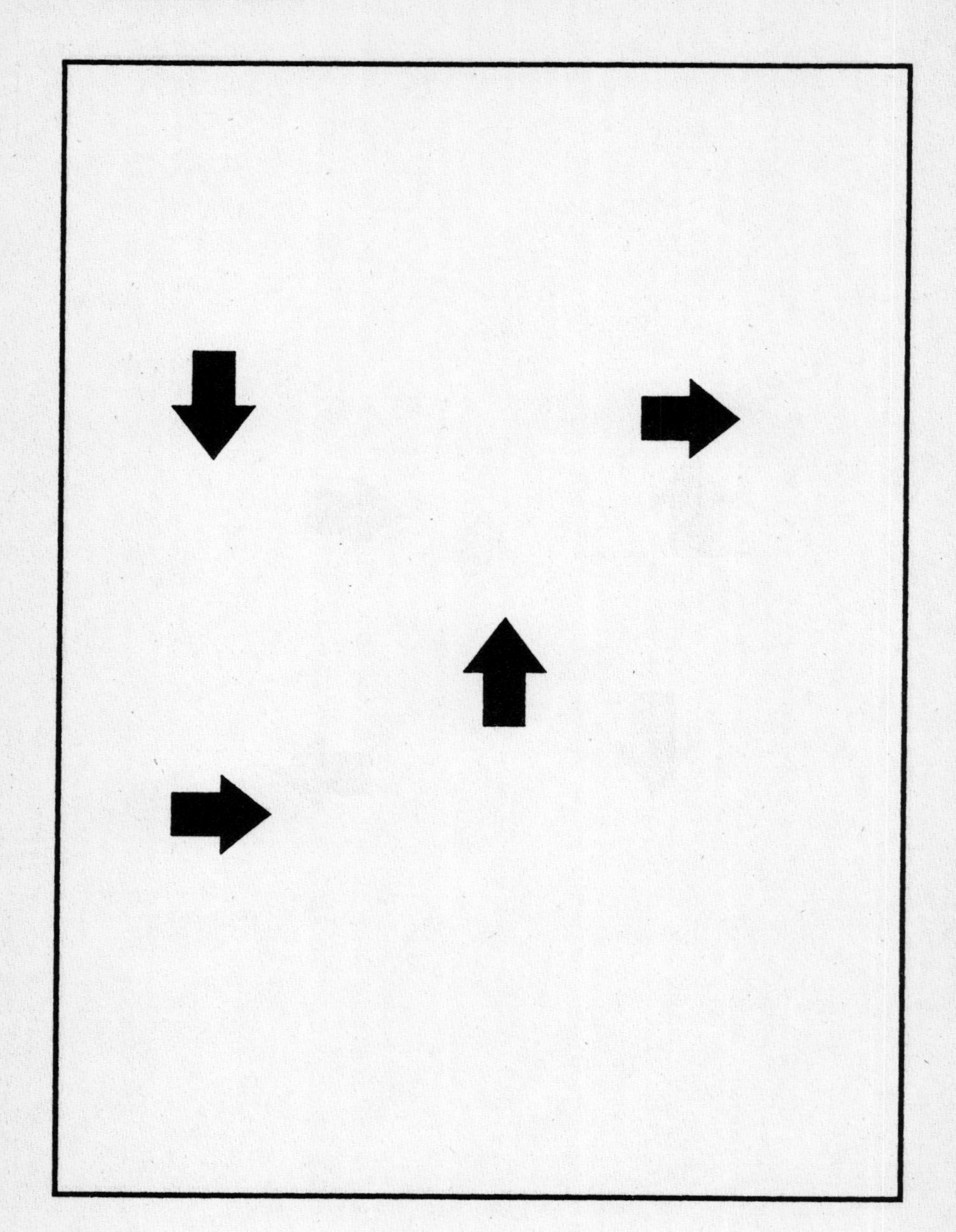

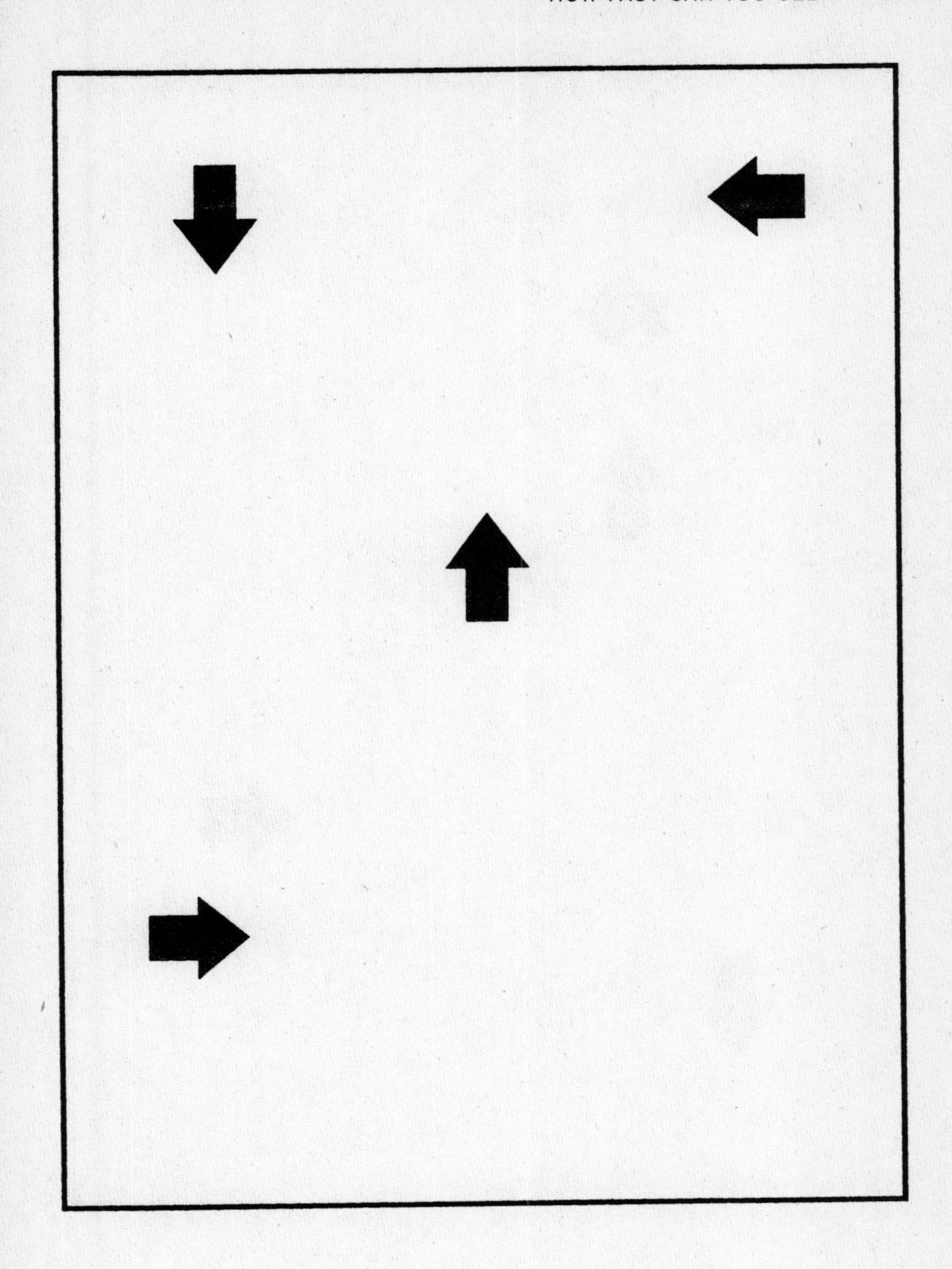

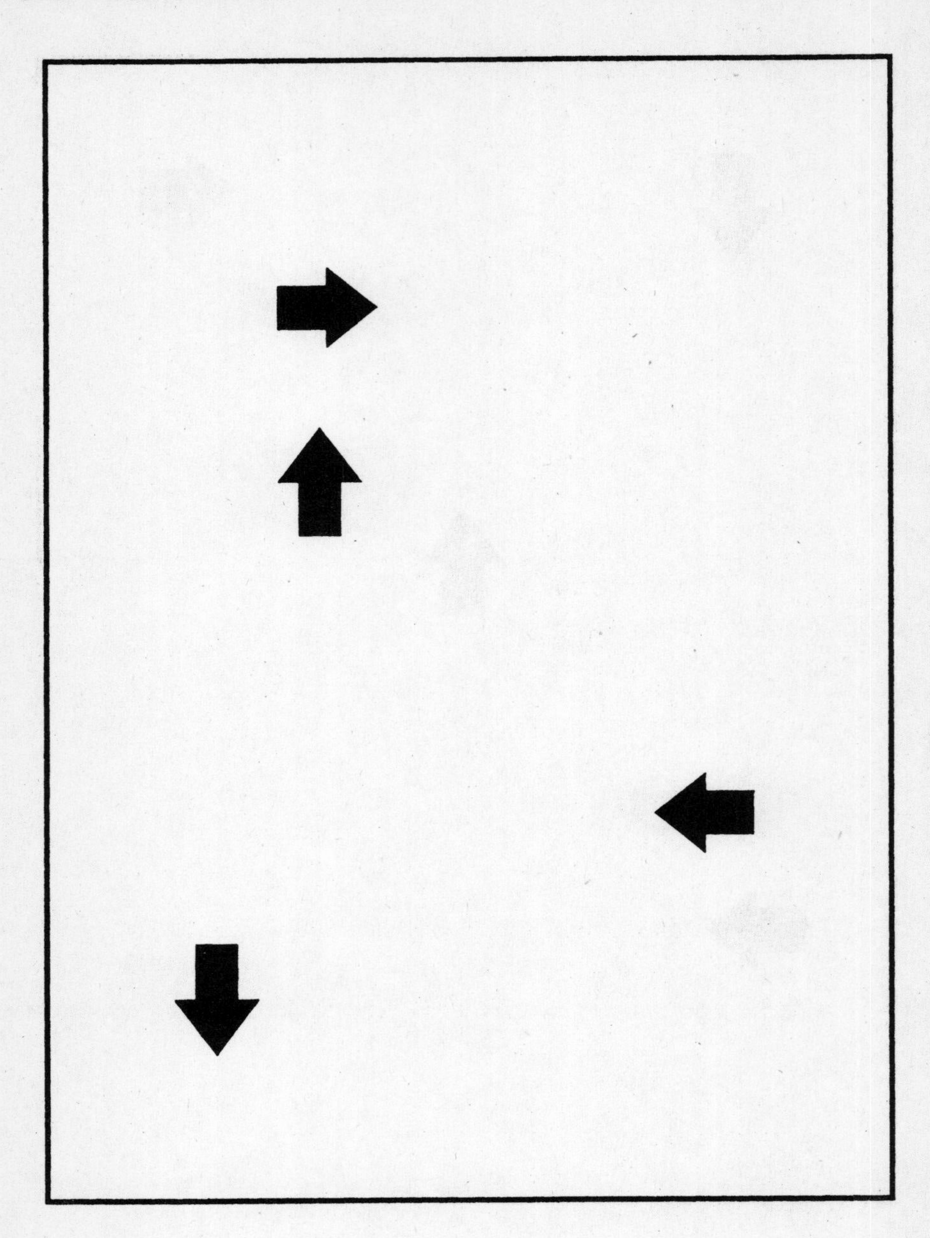

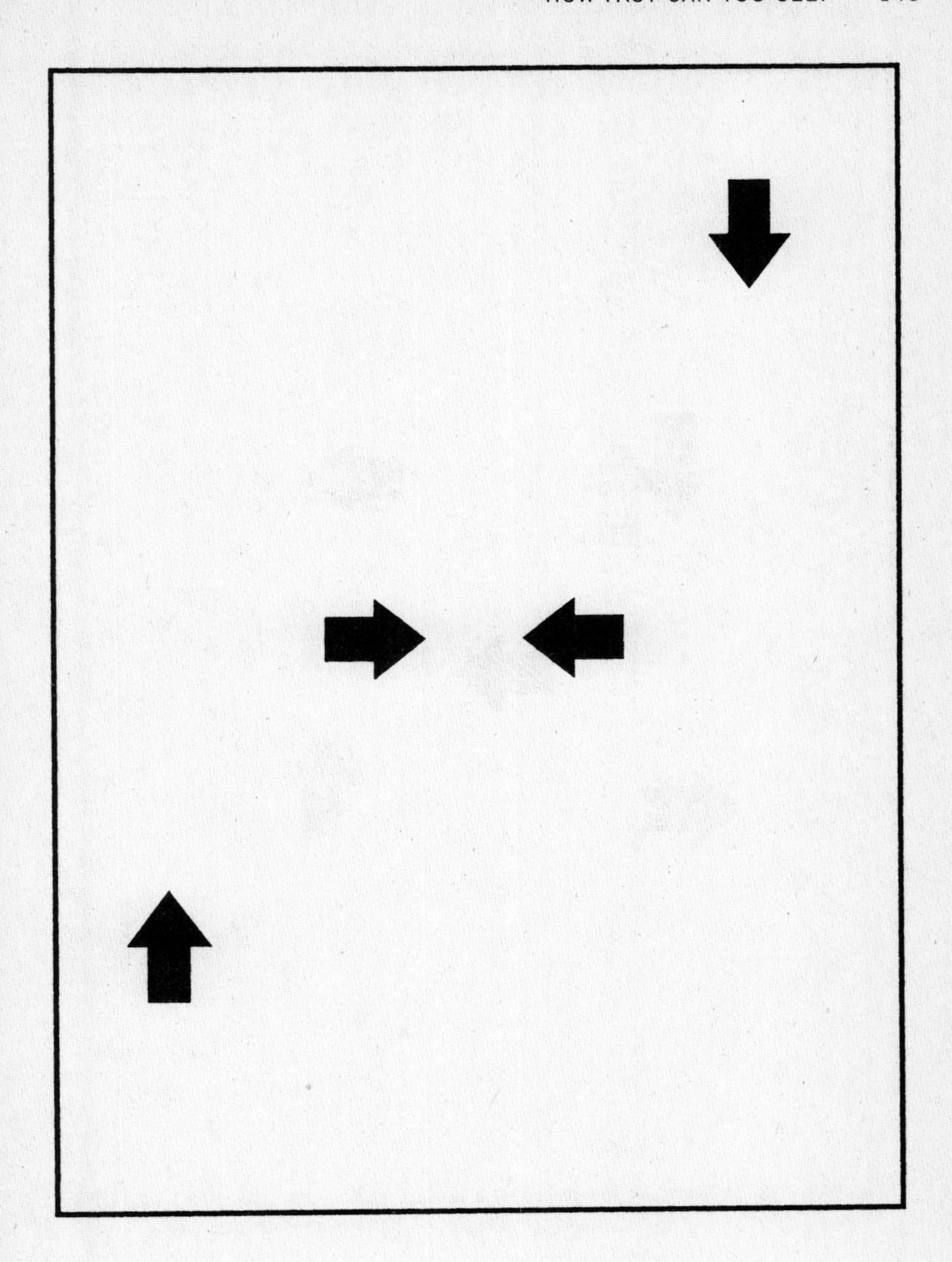

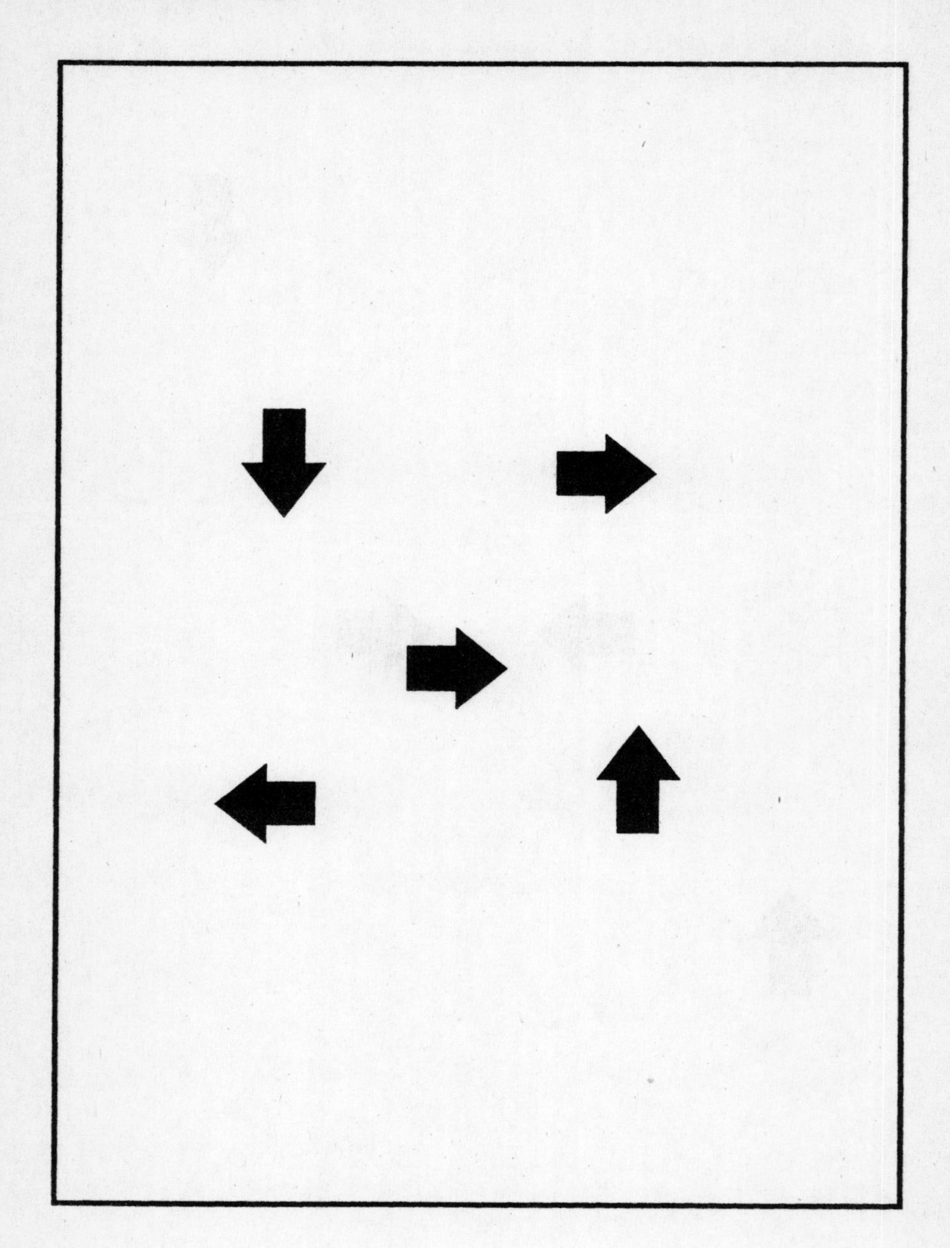

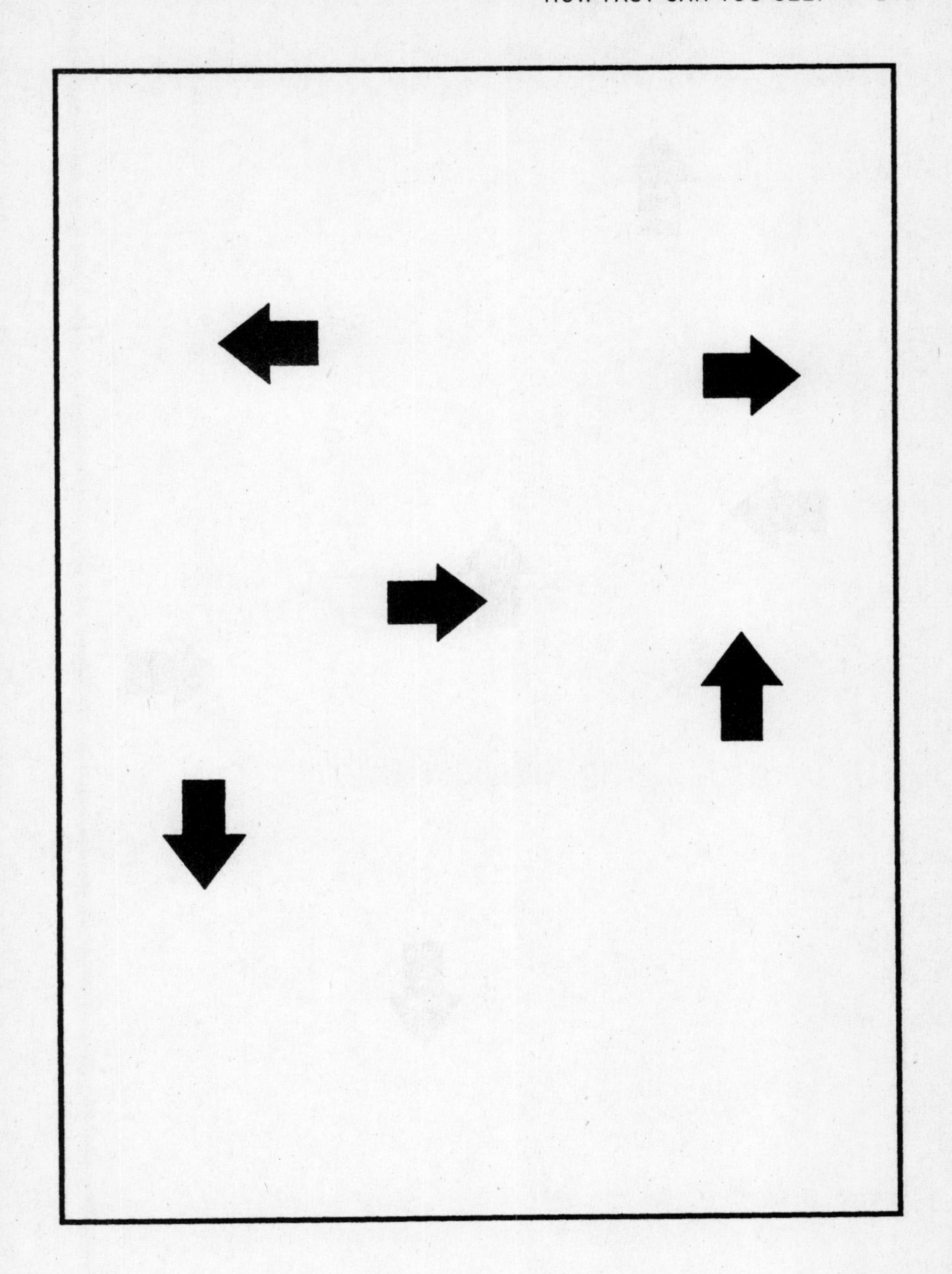

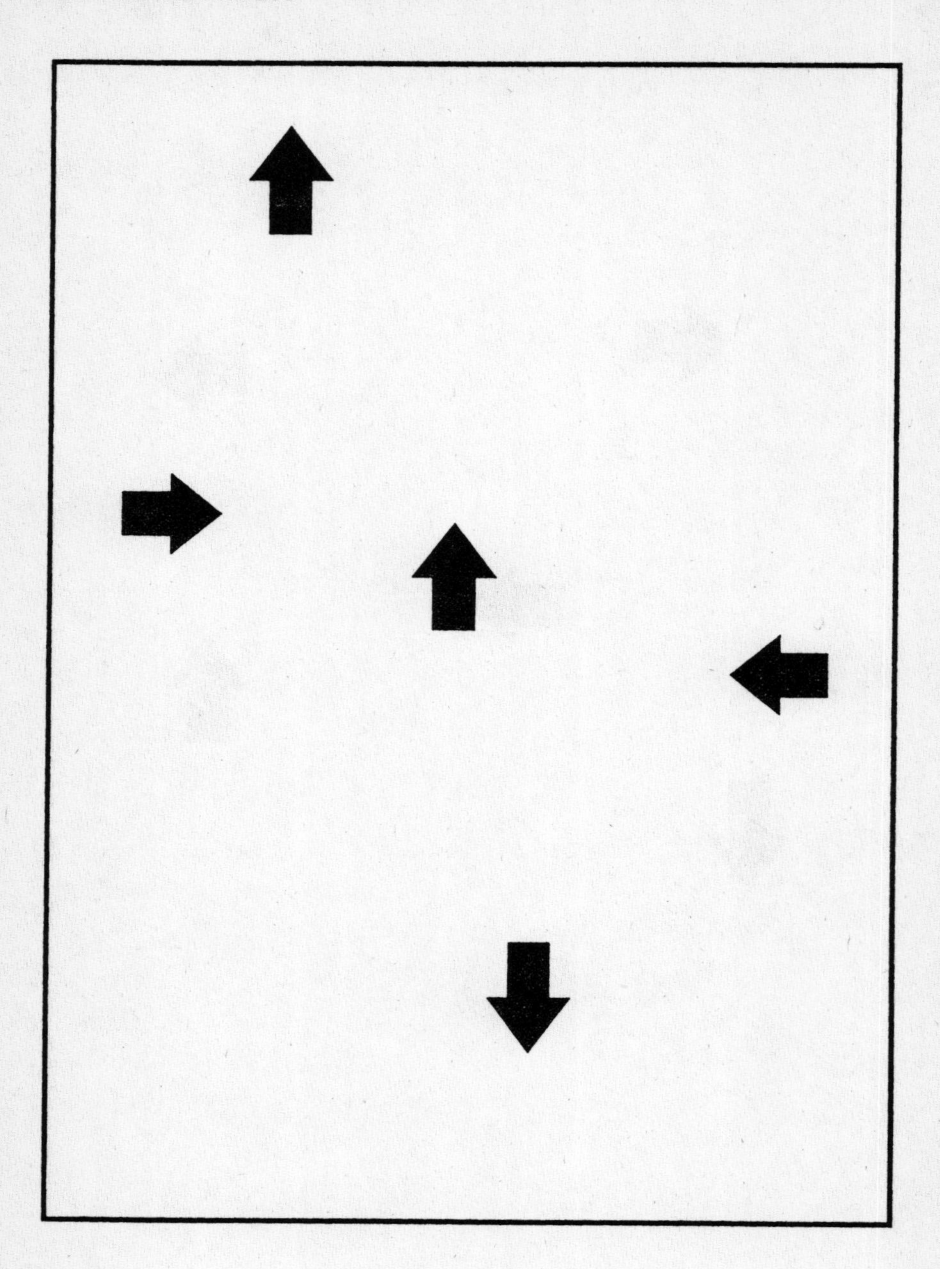

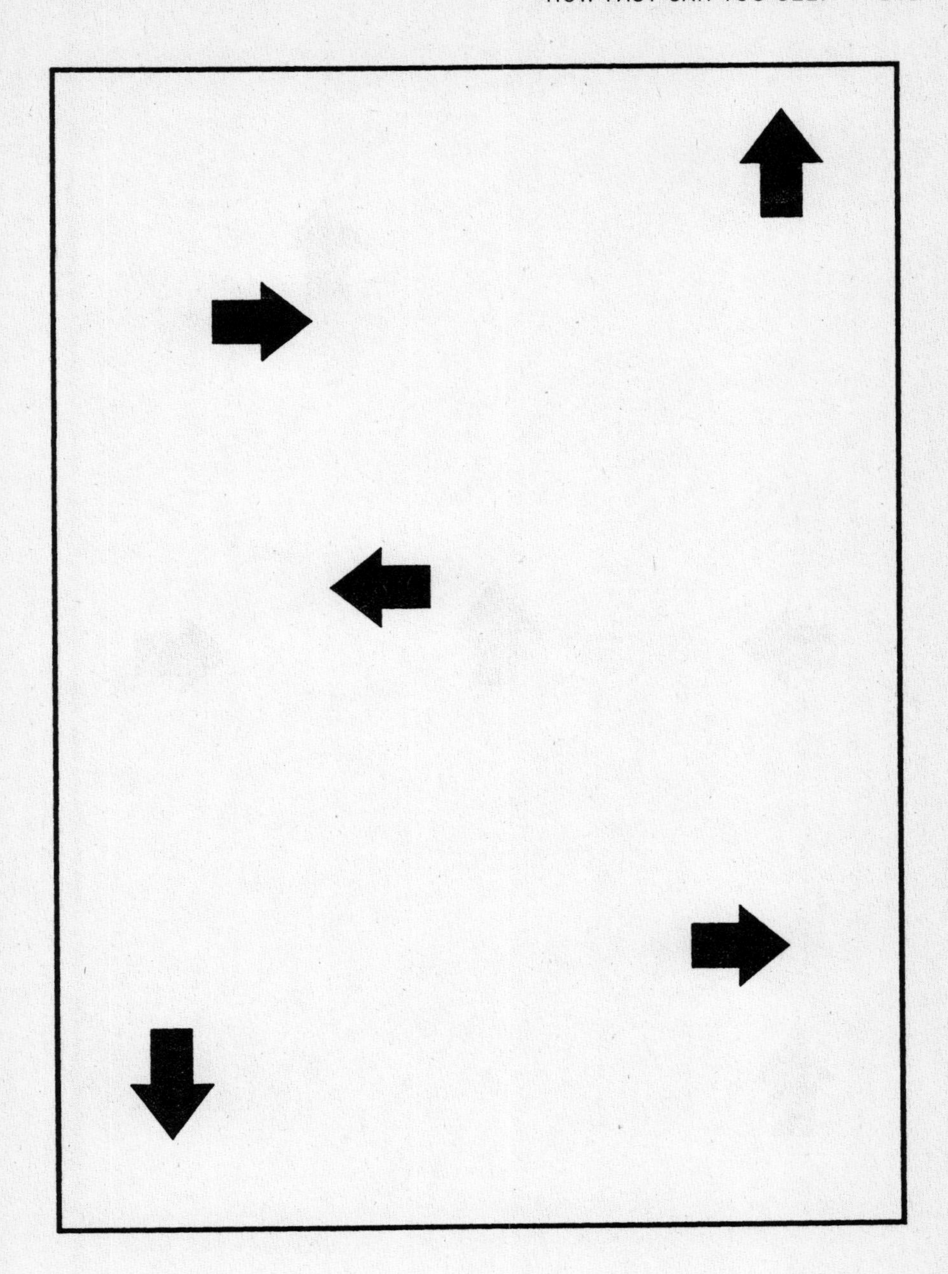

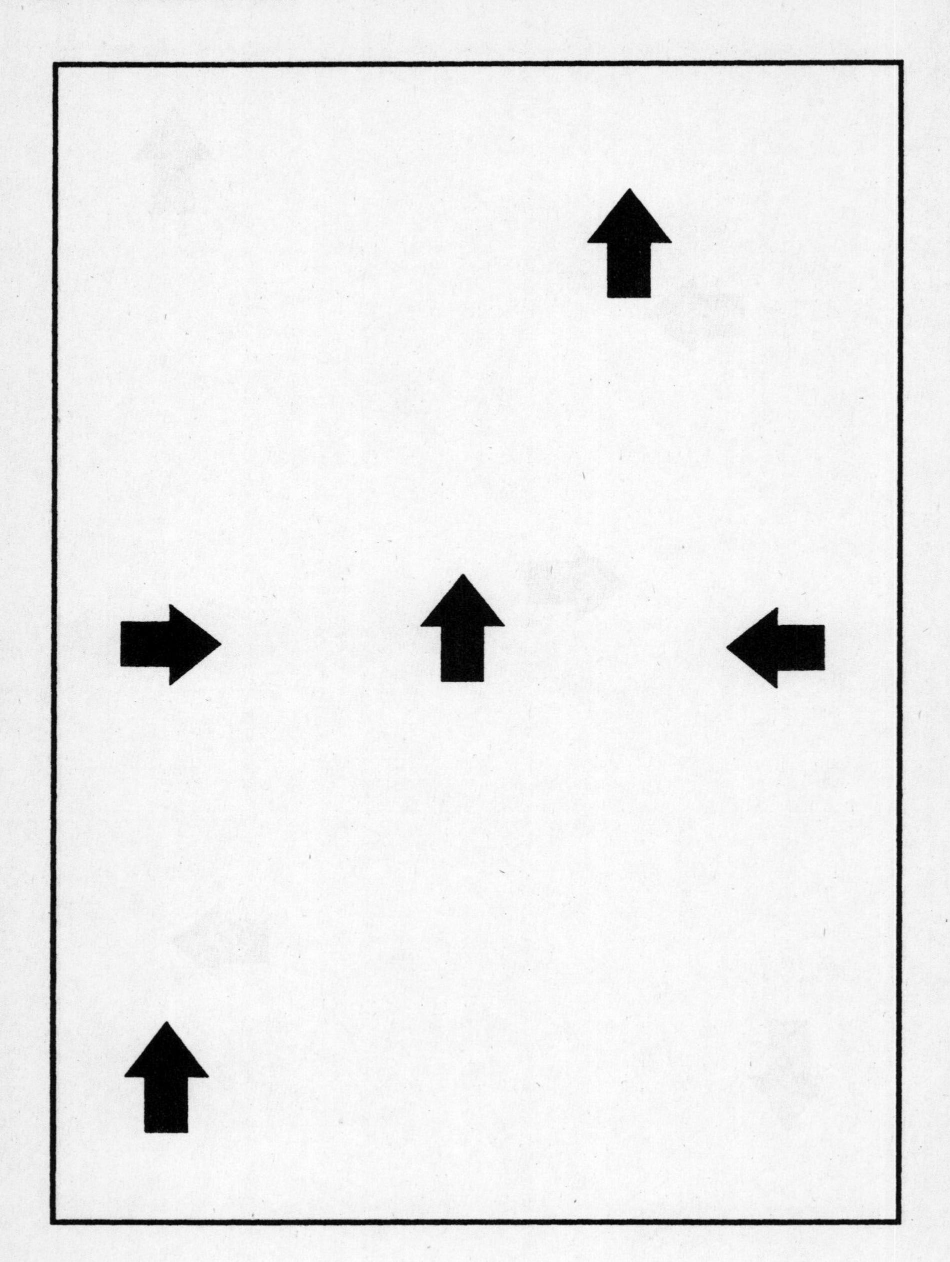

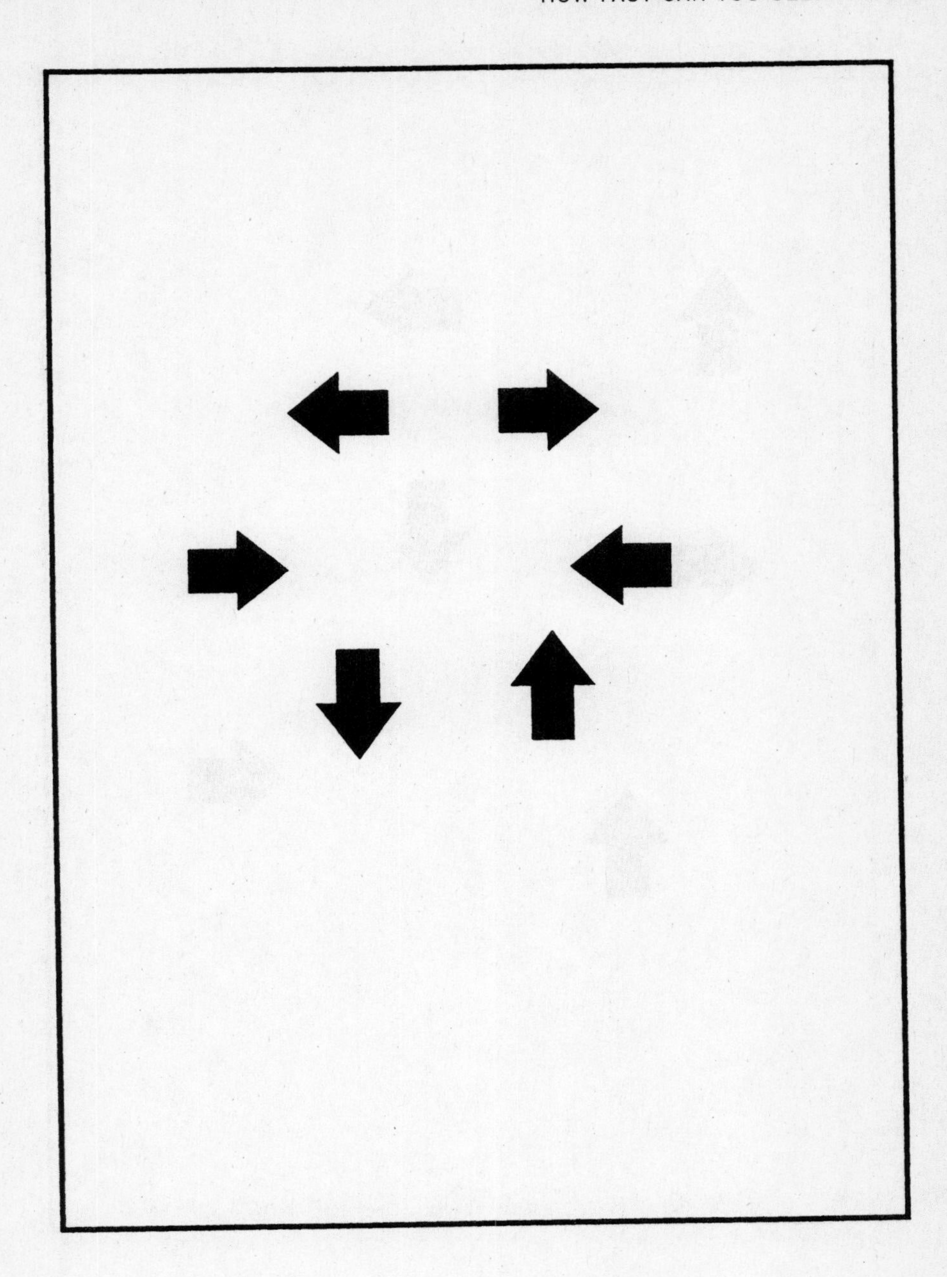

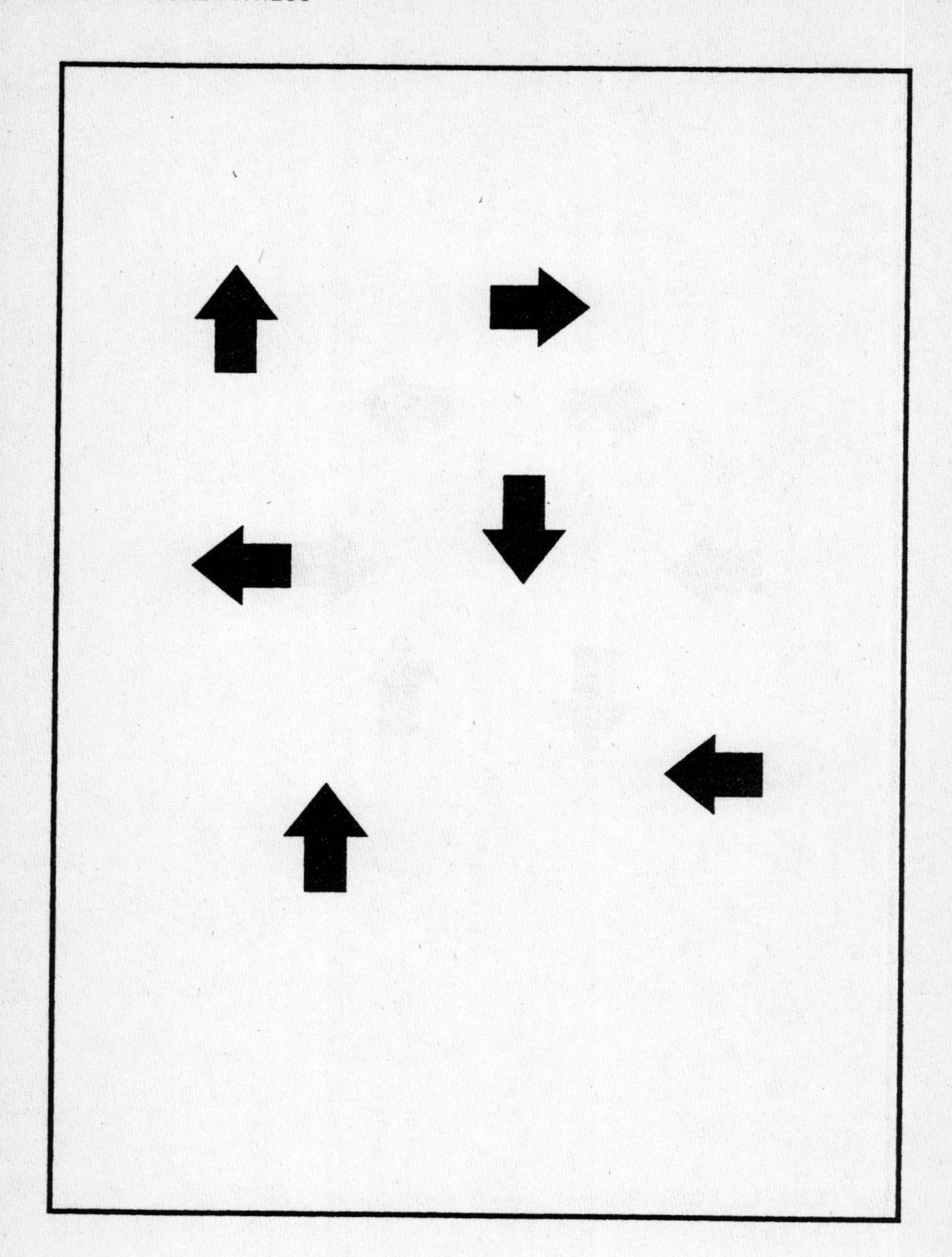

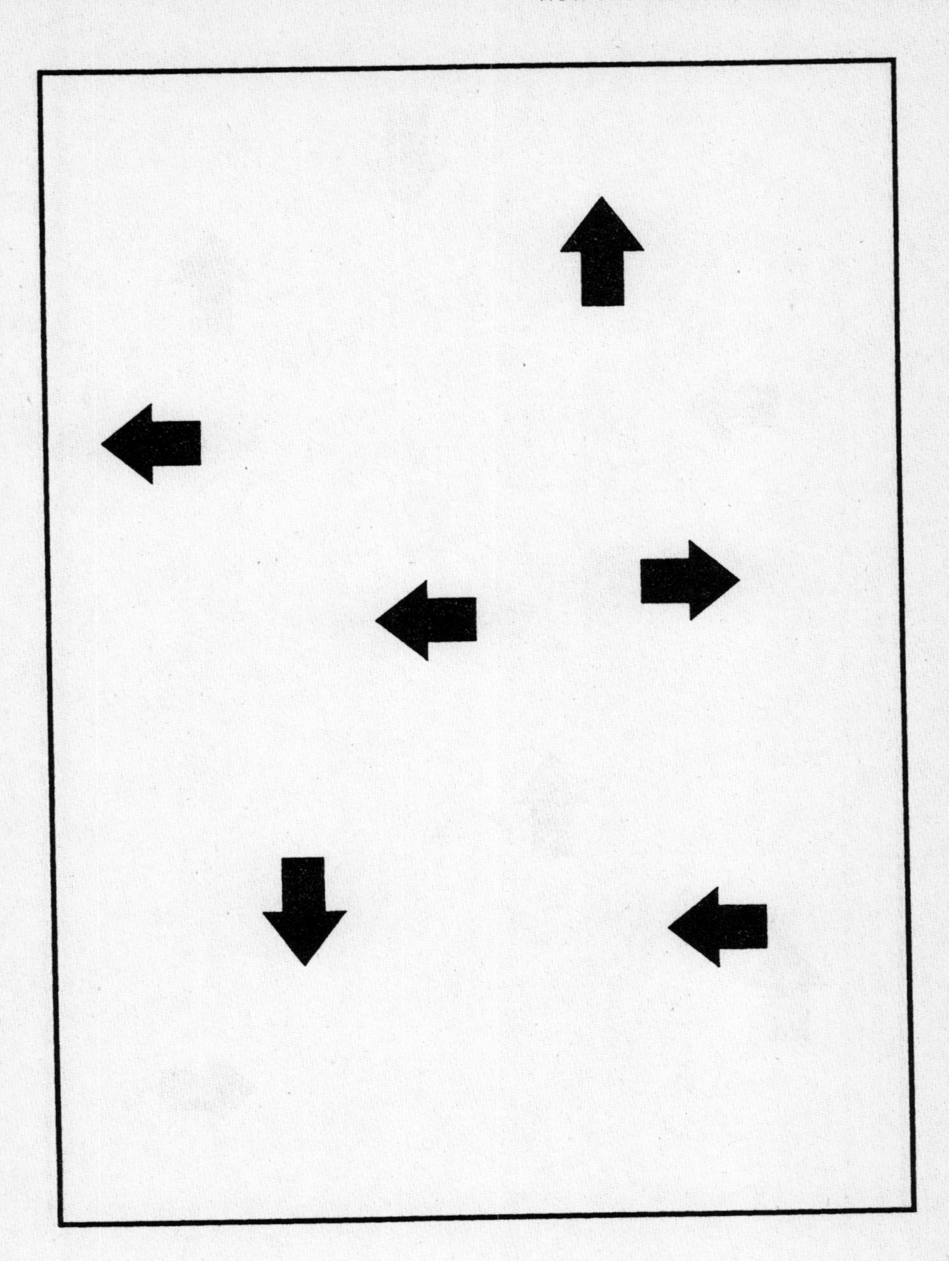

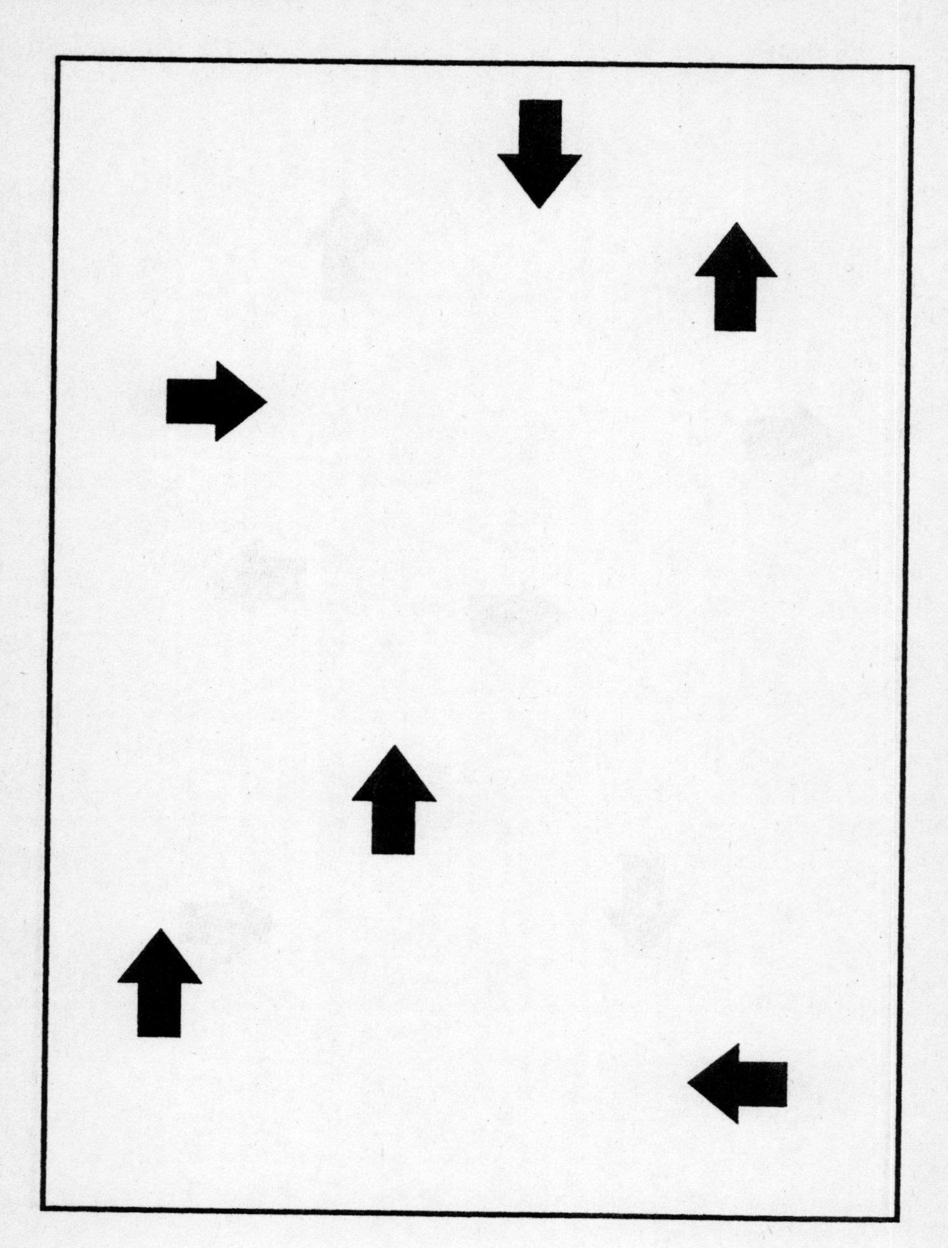

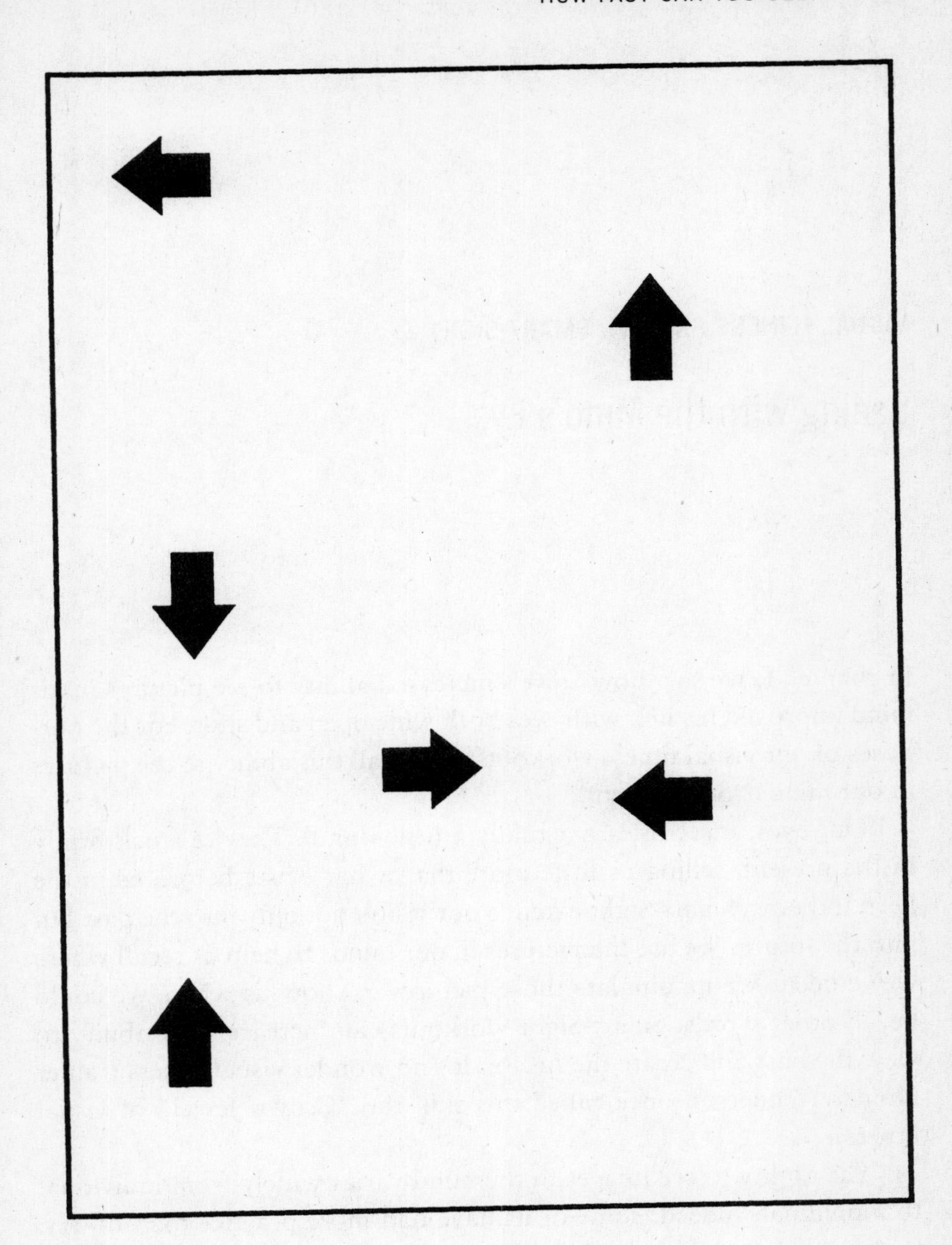

8

VISUAL FITNESS SKILL VI: SMART-SIGHT

Seeing with the Mind's Eye

In chapter 1, we saw how Steve's increased ability to see pictures in his mind improved his life, with eyes both wide open and shut. For the purposes of our visual fitness workouts, we'll call this ability to see pictures in our minds "Smart-Sight."

Our eyes, it seems, are actually a little dumb. They're stuck firmly in the present, telling us little more than what exists before us in the here and now. Smart-Sight extends our vision not only into the past but into the future. We use the pictures in our minds to help us recall where we've been. We manipulate those pictures to show us where we could be. In other words, Smart-Sight workouts can increase your ability to view the past and create the future. It's no wonder visual fitness trainer Homer Henderson once called this skill the "Crown Jewel" of visual fitness.

The ability to see images in the mind varies widely from individual to individual. Indeed, some of us have had more practice than others. In school, for instance, while many of us prepared for the future by completing essays on important subjects such as the nesting habits of birds in Boca Raton, others spent their time staring off into space, honing their visualization skills so that one day they might distinguish themselves as inventors, artists, business owners, or top-level scientists. Einstein is a good example of one of these daydreamers. This less than exemplary student later made a name for himself despite having been suspected of retardation by his more verbally inclined instructors.

Traditional schools center on verbal rather than visual thinking. As a result, our Smart-Sight muscles tend to grow flabby. Most of the adults I examine—other than the architects and artistic sorts—score little better on their Smart-Sight tests than the average ten- or twelve-year-old.

We'll be dividing our Smart-Sight workouts into two sets of exercises: those that increase our ability to recall the past, and those that increase our ability to create the future. The ability to recall the past, using pictures in our minds, we call *visual memory.* The ability to create or manipulate the pictures in our minds—and therefore create the dreams that could guide our futures—we call *visualization.*

Visual Memory

If you've already started on your Quick-Sight exercises, and you're actually picturing the numbers and arrows in your mind, then you've already begun working your visual memory. But what are some of the ways in which visual memory can make our sight smarter?

Spelling

There are three main ways in which we spell, or fail to spell. One way is using a knowledge of the sounds of letters and letter combinations. We call this process phonics. While phonics is the best way to figure out unfamiliar words when reading, phonetic spellers have their problems. The letters *ur, er,* and *ir,* for instance, all make the same sound, but you'll get into trouble if you substitute *er* or *ir* for the *ur* in the word *church.* To make matters worse, words such as *rendezvous* defy the rules of phonics.

A second, more reliable way to spell is to use sequential memory;

that is, you memorize the word much as you would memorize a poem: from beginning to end. For instance, in spelling *Mississippi,* you simply recite *M-i-s-s-i-s-s-i-p-p-i* until you can hold the sounds of the sequence in your mind. All too often, however, the sounds vanish from the mind as soon as the spelling test is concluded.

For those who are either blessed with or have developed a good visual memory, the easiest way to spell is to see the word in their minds and rattle off what they're seeing. What could be simpler? Interestingly, I commonly see students who have excellent visual memory but are still poor spellers. That someone has a skill apparently does not guarantee he'll use the skill. Children encouraged to spell using letter sounds may continue to substitute *k-a-t* for *c-a-t* despite having excellent visual memory. On the other hand, asking spellers to picture their words in their minds will not work for those whose Smart-Sight muscles are still flabby.

Navigation

Another way in which visual memory can enhance our lives is in finding our way home after a visit to a new location. If you can see in your mind where you've been, you have a built-in road map to take you home. Otherwise, like Hansel and Gretel—who obviously shirked their Smart-Sight exercises—you'll need to mark your path home with a trail of bread crumbs.

Learning from the Past

Visual memory—or the lack thereof—may affect our lives in unsuspected ways. Billy Wilder once said that hindsight is always 20/20. Of course, Wilder was a screenwriter and director, not a visual fitness coach, and the truth is, in the world outside the silver screen, hindsight can fall miserably short of 20/20. All too often, we find those whose hindsight plays more like blind-sight. Such individuals seem incapable of learning from their mistakes.

The best way to learn from the past is to be able to see the past, and the best way to see the past is a good Smart-Sight workout.

Visualization

Smart-Sight, however, includes more than just hindsight. We also have to consider foresight. While hindsight depends on visual memory, foresight depends on visualization. The ability to visualize is the ability to see in your mind something that does not yet exist, something that is just a possibility. It is what you do when you imagine yourself crossing the finish line in an upcoming race; it is what you do when someone describes a new dress they bought, and you're able to see the dress in your head.

Navigation

Let's return to navigation, but this time from the viewpoint of visualization rather than visual memory. Just as visual memory is useful in recalling where we've been and how to retrace our steps, visualization is useful in guiding us to new locations. Good visualizers can generally find their way with a general idea of where they are going. Give them a peek at a map of Manhattan, tell them that their destination is on Fifth Avenue a couple blocks south of the Metropolitan Museum, and they'll be on their way. Whether they amble north on First, Second, or Third Avenue, and cross over on Fifty-seventh or Sixty-second Street, they could care less. Their destination is in their minds. Exactly how they get there is of little importance.

Those who can't visualize need more help to feel comfortable in unfamiliar surroundings. For instance, you might have to tell them, "Go north on First Avenue. You should pass Fifty-seventh Street, then Fifty-eighth. If you come to Fifty-sixth Street instead, you're going south. Turn around and go back in the opposite direction." And so forth. Those who can't visualize are more likely to become lost when they visit new places.

Thinking, Reading, and Writing

Some people naturally think in images; they see pictures in their minds. We call these people visualizers. Other people have trouble creating mental images. Instead they rely on verbal chatter in their thought process. We call them verbalizers. Let's go back to the example of a new dress to illustrate the difference between these thinking styles. Suppose I wanted someone to shop for a certain dress for my wife at a particular store. If I described the dress to a visualizer, she would go into the store and simply find the dress that matched the image in her mind. A verbalizer would use a different process. He would go into the store, pick up a dress, and think something like this: "David said the dress had short sleeves. This one has long sleeves. David said the dress had a blue collar. The collar of this dress is red. This is not the same dress." Of course, the verbalizer's thought process might be much quicker than this—almost instantaneous—but this is the process he would go through.

Our style of thinking affects more than the way we shop. Whether we are visualizers or verbalizers can also influence how we'll read and write. Consider the following:

> As the rain stung, forked lightning streaked the gunmetal gray of the sky. Waves swelled and whitecaps frothed while the tiny boat rose and fell, its shredded sail flapping, its mast rocking side to side as if waving goodbye.

Can you see the images in your mind: the lightning, the sail, the rocking boat?

Now consider this second passage:

> Shocked and repulsed, I reread the hateful message. His words were frightening, horrifying actually. How could anyone be so cruel? How could anyone be so evil?

What type of images does this passage elicit? Quite a difference, huh?

The first passage paints a concrete, visual world of pictures for the reader. The second uses verbal abstractions such as "frightening," "horrifying," and "evil" to evoke emotions and ideas, not pictures.

Now, none of this is to say that ideas are inferior to images. The point is, our visual style—whether we visualize or we verbalize—can spill over into the way we write. And think.

In the same way, our visual style can affect our reading habits. If, for instance, when you read the first passage, you had no picture in your mind of the gray sky, the waves, or the boat, it's less likely that the passage would have appealed to you. True, you might have enjoyed the sound of the words, but reading without pictures in your mind is a little like seeing *Star Wars* with your eyes shut.

The ability to picture what we're reading allows us to "be there," whether "there" is stumbling along a cracked sidewalk bordered by a dead, brown lawn or "there" is swinging from a stout vine in a lush green African jungle. The child—or adult—who enjoys watching stories on TV but wouldn't consider listening to a book on tape may be experiencing a difficulty with visualization.

Sports

In 1974, I was cocaptain of the U.C.L.A. Gymnastic Team. When I look back to that time from the perspective of a visual fitness coach rather than a gymnast, it's now clear to me that the moves I could throw in my mind were the moves I could indeed perform. The moves I couldn't see in my mind, I never really mastered.

The role of visualization in sports is probably best suggested by a well-known experiment that I learned of in professional school. An Australian researcher named Alan Richardson conducted the experiment, which explored the effect of visualization on performance in basketball. He took three groups of students and tested their ability to make free throws. Afterward he had one group do nothing concerning basketball, one group practice shooting baskets daily, and one group practice visualizing themselves shooting baskets daily. The visualizing group shot no actual baskets. Instead, they just pictured their bodies making the shots, pictured the ball arching through the rim.

At the end of twenty days, Richardson retested his students. As would be expected, the group who had not concerned themselves with basketball showed no improvement. Less predictably, the group who shot baskets only in their heads showed almost as much improvement as those who had practiced the actual shooting of baskets: the visualizers improved 23 percent, the shooters 24 percent.

In this chapter we aren't going to discuss the application of visualization to any sport. We are, however, going to help enhance the skills you'll need before you can take full advantage of your golf or tennis professional's suggestions.

Creating a New World

Most motivational speakers would agree that the first step in attaining a goal is to see that goal. The lightbulb turned on in Thomas Edison's mind long before it turned on in the homes of our nation. That nation, in turn, glowed in the minds of our founding fathers long before there was any United States.

In previous chapters, we've explored how our eyes bring the world into our minds, but visualization allows us to create in our minds pictures that we can then bring into the world. Airplanes, moon walks, and the Internet, not to mention Robinson Crusoe and Harry Potter, all appeared as images behind eyes long before they appeared as images before eyes. How many of the innovators in business, science, and the arts pictured success long before they obtained it? Much of the world's comfort, prosperity, and beauty is due to these innovators' ability to visualize.

Self-improvement gurus tell us to visualize goals, visualize success, visualize our golf games, not to mention world peace. If we fall ill, they even ask us to visualize the body's immune system devouring any invading viruses or cancer cells. These techniques may be great for those who are already adept at visualizing, but what about those with poor visualization skills? The Smart-Sight exercises in this chapter will help anyone begin to create pictures in their mind's eye.

Stuck in Your Head

As defined above, visualization is the ability to create, control, and manipulate the pictures in your mind. That includes, one might add, the ability to make those pictures vanish. The daydreamer who can't stop daydreaming, the accident victim who can't help seeing the accident, the absentminded professor lost in his or her head are all examples of people who are controlled by their mental images rather than controlling those images. Just as you should have the ability to see either big or little, you should have the control to see either inside or out. The workout that follows, added to the workouts you began in previous chapters, should move you in the direction of such control.

Your Smart-Sight Workout

Between individuals, there is a wide variety in Smart-Sight muscle development. At one end of the spectrum, for instance, we have those whose images are eidetic; that is, their mental pictures are so exact that such persons are said to possess a photographic memory. At the other extreme we have those who are seemingly devoid of the ability to recall or generate images at all. Most of us fall somewhere in between these two extremes and, with practice, we can inch our abilities at least a little further in the photographic direction.

If we can return for a moment to Richardson's basketball visualization experiment, there's a key point we didn't cover earlier. When Richardson interviewed his participants, he found that it was their ability to manipulate the images in their mind, not how realistic their images were, that was most related to the level of their improvement. This observation supports the idea that the precise recall of images, while a major help to memorization work in school, may be less important to changing the future than our ability to create, control, and manipulate those images in our minds.

In chapter 7, the Quick-Sight chapter, we worked with flashing numbers and arrows. In addition to working the speed of perception, these drills were also designed to develop visual memory—providing that you maintained the image of the numbers or arrows in your mind. The present chapter works with visualization rather than visual memory. While you go through this Smart-Sight portion of your workout, you will, of course, continue to work Quick-Sight I, II, and III from the previous chapter. As you work these Quick-Sight exercises, be sure to see the numbers and arrows in your mind. If you have difficulty holding those pictures in your mind, however, the following Smart-Sight exercises may be just the thing you need.

Smart-Sight I

In the chart on page 165, there are two arrows, a tiny one at the top and a large one at the bottom. For this drill we will only be using one arrow. Some people have an easier time with the small arrow, others with the larger arrow. Try the drill both ways to see which size works best for you. Use a piece of paper to cover the other arrow so only one is visible at a time.

Here are the exercise steps. You'll find them easier if you have a partner who walks you through them.

1. Relax and open up your side vision as in the Big-Sight exercises. Look at the arrow. Close your eyes. Try to continue relaxing and to see the arrow in your mind.

   Okay now. Don't get upset. Most people do not see an arrow in their minds that is an exact replica of the real arrow. Their mental images are often vague. Picture your bedroom, for instance. Where are the bed, the dresser, the door, and the windows? The image in your mind might not run quite like a video from *Lifestyles of the Rich and Famous*. The images may look more like those in a dream,

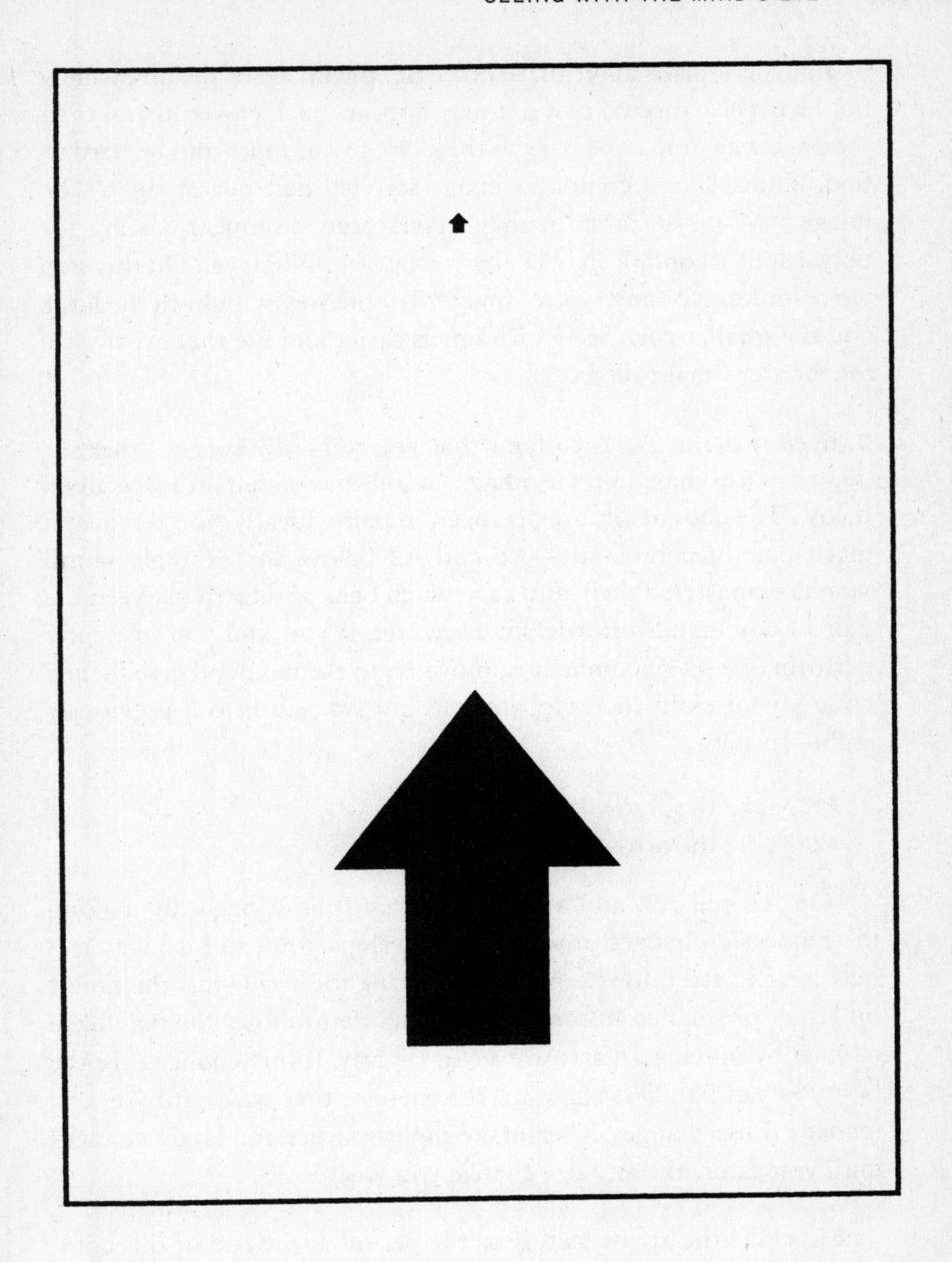

or a daydream. To make matters worse, the image of the arrow may not be in the expected place; it may appear much closer to you than the page was. It may be over to the side of your mind and be hard to find, almost like a computer image scrolled half out of sight. The image may be unstable; it may appear and disappear, lasting for only a split second. It may be the wrong size. Whatever. On this step we're looking for any type of image. Try the step with both the large and the small arrow. See which one is easier and use that arrow size for the steps that follow.

2. Even if the image is so faint that you're hardly sure it's there at all, practice changing the image in the ways that are described below. The commands are arranged in pairs. Ideally, you'll repeat a given pair of commands—A1 and A2 below, for example—until you have mastered them and can switch back and forth between the pair of commands effortlessly. However, if you find you just can't perform one set of commands, move on to the next and give them a try. As your skills increase, you can always return to a previously stubborn step.

A1. Make the arrow larger.
A2. Make the arrow smaller.

On A1 and A2, and all the pairs that follow, begin by making the requested changes in whatever size steps you can manage. You may have to start, for example, by making the arrow just the tiniest bit larger or smaller. In some of the pairs that follow, you may have to start by moving an arrow ever so slightly. If tiny changes are the best you can handle at first, do the exercise that way until you can manage those changes. Then make slightly larger and larger changes until you can make any size change you wish.

B1. Make the arrow as tiny as the period at the end of this command.
B2. Make the arrow as big as a wall.
C1. Tense up and make the arrow disappear.

C2. Relax and make the arrow reappear.
D1. Move the arrow closer to you.
D2. Move the arrow farther away from you.
E1. Move the arrow farther to the right.
E2. Move the arrow farther to the left.
F1. Move the arrow higher.
F2. Move the arrow lower.
G1. Flip the arrow so it points down.
G2. Flip the arrow so it points up.
H1. Rotate the arrow so it points to the right.
H2. Rotate the arrow so it points to the left.
I1. Rotate the arrow clockwise so that it points up, right, down, left, up.
I2. Rotate the arrow counterclockwise so it points up, left, down, right, up.
J1. Repeat the D steps through the I steps. After each step, tense up and make the arrow disappear.
J2. Relax and make the arrow reappear.

3. If while you are trying the above steps you cannot see any image, then try one or more of the following alterations:

A. Be sure you are relaxed. Remember to breathe. With your eyes closed pretend you are seeing your bedroom. If you're able to see any image at all in your mind, manipulate that image rather than the arrow.
B. If you're still getting nowhere, make the arrow as tiny as possible in your mind, little more than a speck. Increase the size of the arrow as your ability to visualize improves.
C. If that doesn't help, close your eyes and be aware of the darkness. Make the arrow the same shade of darkness so that even though you can't see the arrow, you *know* the arrow is changing as in the above paired steps A through I.
D. After working this way for a time, try to turn the arrow white or gray so that it stands out in the darkness. Alternate between the arrow being the same shade as the darkness and it being

some other shade of darkness: black, dark gray, brown, dark red, whatever. This step will probably be easier if you make the arrow as small as possible—the size of a pinhead if necessary—then gradually increase the arrow's size as you are successful.

E. If you can't do step D, forget about the arrow entirely. Close your eyes. Gently cover them with the palms of your hands. Look right through the darkness. Try to change some part of the darkness, at any distance in any way. Move some part of the darkness closer or farther or to the side. Change the color of some part of the darkness—any part, large or small—to bright yellow or blue or red or orange, any color you can produce in your mind, even if only for an instant. Then make the color vanish and return to the original darkness. Try to make tiny changes. Try to make large changes. Your goal on this step is to change anything at all in your mind—the position, the color, the shade, anything. After you've made a change, uncover your eyes and look across the room at some solid object. Then close and cover your eyes and try to make a change in the darkness.

If picturing things in your mind remains difficult, don't get discouraged, just put in your seven minutes and move on. As you continue to work your other visual fitness muscles, such as Clear-Sight and Big-Sight, you may find you have an easier time seeing pictures in your mind. Sometimes learning to increase the control and flexibility of your outward seeing can also improve your inward seeing.

Smart-Sight II

In this drill you will be using the same arrows you are using in Quick-Sight II. Turn to the chart on page 124. You will now, start to learn to manipulate those arrows in your mind. To do this exercise, you'll need a pencil and paper.

90-DEGREE ROTATION

In this section, as explained below, you will be rotating the arrows in your mind by 90 degrees. Perform the following steps:

1. Start with a line of three arrows. Flash them as you learned to do in Quick-Sight II.

2. When you have covered the arrows back up, picture them in your mind as if the line were the hand of a clock pointed to the *3* on the clock. Repeat steps 1 and 2 as many times as it takes to get a picture of all three arrows in your mind. If you can't keep the arrows in your mind, leave them uncovered.

3. Now, in your mind, rotate that line of arrows 90 degrees counterclockwise until the hand on the clock is pointed to the *12*.

If the three arrows were initially pointing up, they would now be pointing to the left as in the picture below. If the arrows initially pointed up, down, and right, they would now be pointing (from the bottom of the line to the top) left, right, and up.

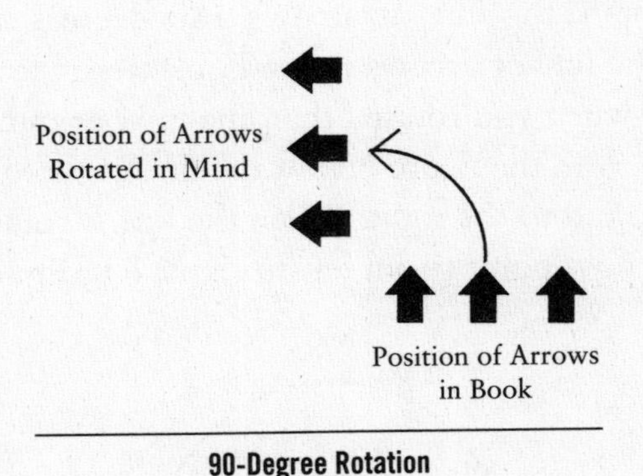

90-Degree Rotation

4. Draw what you are seeing in your mind. Uncover the arrows and rotate the book 90 degrees counterclockwise, just as you rotated the line of arrows in your mind. (The top of the book will now be to your left.) Compare the vertical line of arrows in the book to

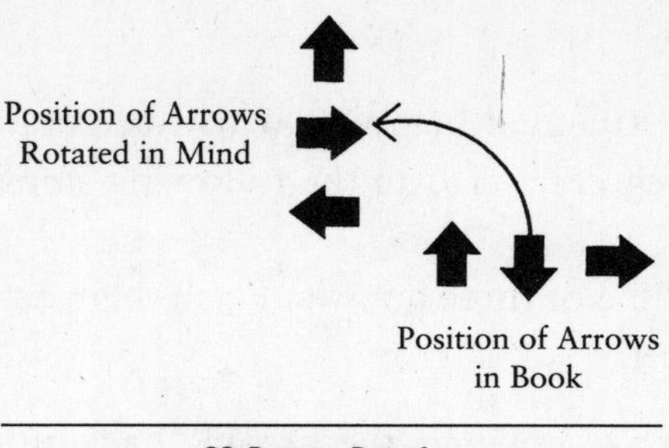

90-Degree Rotation

what you drew. If you manipulated the arrows correctly in your mind, the book arrows and your drawing should match.

5. Work through all twelve three-arrow patterns—always beginning with the book in its upright position—until you can easily rotate each pattern 90 degrees.

180-DEGREE ROTATION

In this section you will repeat the same steps of the last section, but this time you will rotate the line of arrows 180 degrees at each turn. In other words, if the line of arrows in the book were the hand of a clock pointed at the *3,* when you rotated that line in your mind it would now be pointed at the *9.* In the figure below, the line of three arrows initially points up, down, and to the right. When the line is rotated 180 degrees the three arrows now point (from left to right) left, up, and down.

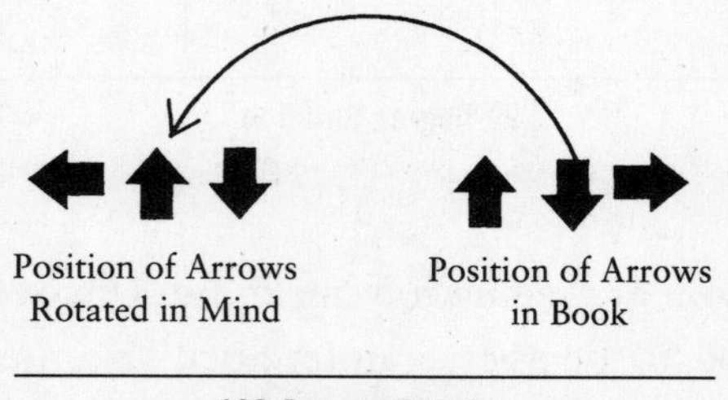

180-Degree Rotation

Again, work through all twelve three-arrow patterns, until you can easily rotate each in your mind 180 degrees and draw them in this position. Then move on to the 270-degree rotations.

270-DEGREE ROTATION

In this exercise, if the three arrows formed the hand of a clock initially pointed at the *3*, you would rotate the hand counterclockwise in your mind until the line pointed down at the *6*. For example, if the arrows initially pointed up, down, and to the right, they would now point (from top to bottom) right, left, and down.

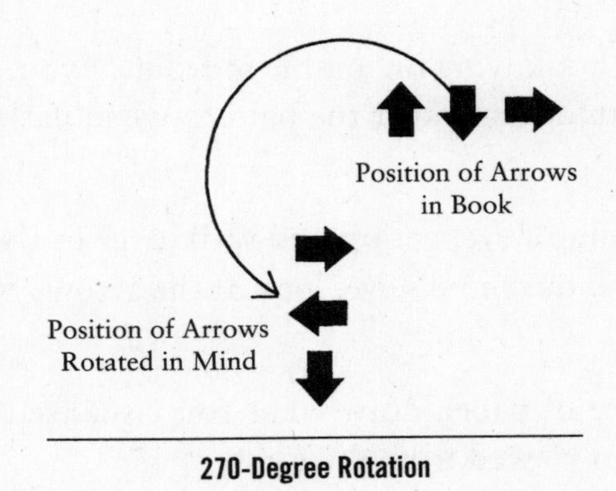

270-Degree Rotation

As before, you will work through all twelve three-arrow patterns, repeating steps 1 through 5, until you can easily flip the patterns in your mind 270 degrees. When you have accomplished this goal, you can begin with the twelve four-arrow patterns on pages 127–129, rotating each 90, 180, and 270 degrees. You will finish by similarly rotating in your mind the five-arrow and six-arrow patterns on pages 130–135.

However, as soon as you can rotate three arrows to all three orientations, you can move on to Smart-Sight III, using Smart-Sight II during one workout and III during the next. If it is difficult to rotate the three-arrow rows, you may want to spend more time on Smart-Sight I.

Smart-Sight III

In Smart-Sight II you're learning to shift the world in your mind. In Smart-Sight III, you're going to learn to shift your point of view: instead of rotating the arrows in your mind, you're going to move yourself in your mind.

For this exercise you'll be using the same arrow patterns that we use in Quick-Sight III on pages 136–155. Follow these steps:

1. With the book lying on a table in front of you, flash the pattern, or if you cannot remember the pattern when flashing it, then study the pattern.

2. In your mind's eye, get up and walk over to the right side of the table. Still in your mind's eye, look at the arrows from the right side of the table.

3. On a piece of paper, draw what you visualized the arrows looking like when viewed from the right.

4. In your mind's eye again, walk to the opposite side of the table from where you were originally sitting. Still in your mind's eye, look at the arrows from there. Draw what you visualized the arrows looking like when viewed from that side of the table. Repeat this step moving to the left side of the table in your mind's eye.

5. To check your accuracy, rotate the top of the book 90 degrees—one quarter turn—to the right (clockwise), so that the top of the book is now to your right. Compare the book arrows to the first pattern you drew. The two should look pretty much alike—including the spacing of the arrows. Keep rotating the book 90 degrees to see how the arrows look from the opposite side of the table and then from the left side.

To see how this would work, look at the three arrows below.

In this figure, we are viewing the arrows as if we are holding the book in the normal position, the bottom of the page closest to us, the page number upright.

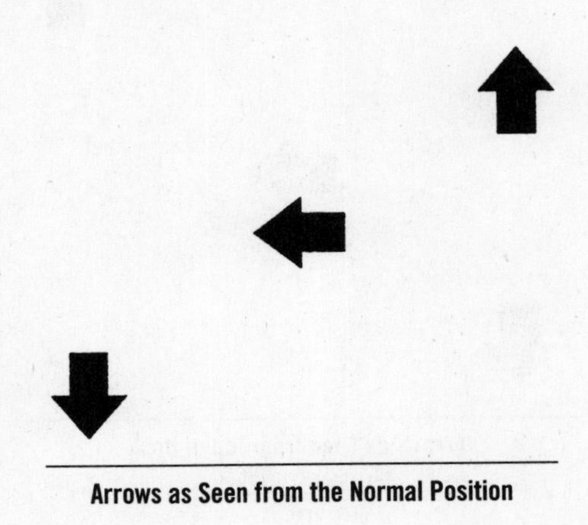

Arrows as Seen from the Normal Position

If we were to walk over and look at the book from its right-hand side, the three arrows would look as they do here:

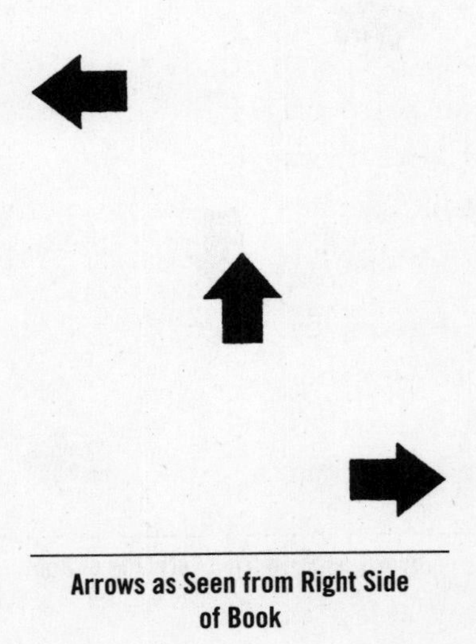

Arrows as Seen from Right Side of Book

If we were to view the arrows by looking from the top of the book, they would look like this:

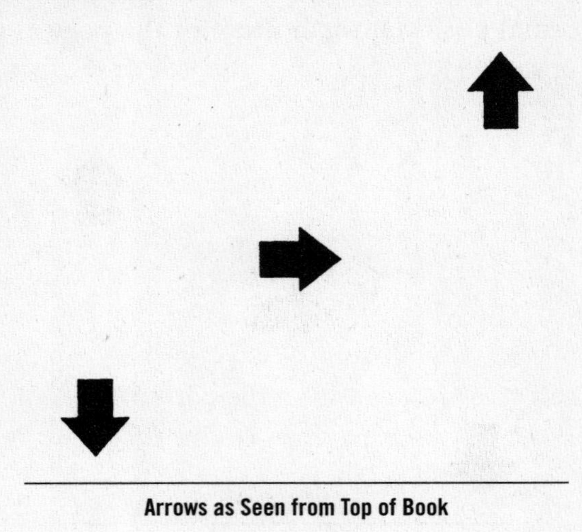

Arrows as Seen from Top of Book

If we were to view the arrows as if by looking from the left side of the book, they would appear as they do in here:

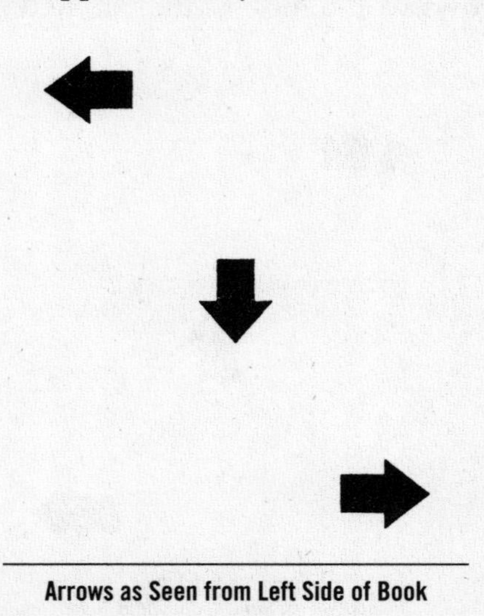

Arrows as Seen from Left Side of Book

In Smart-Sight III you will work your way through the three-arrow, four-arrow, and five-arrow arrangements appearing in charts from the Quick-Sight chapter (pages 136–155). You'll visualize each arrangement as if from the right, top, and left and draw your picture. Then you'll rotate the book and compare to see how you did. To check the pattern as seen from the top, turn the book upside down. To check the pattern seen from the left, turn the top of the book to the left (counterclockwise).

If at first you need to keep the pattern before you when you work, that's okay. You may find it helpful to say the direction of the arrows to yourself to hold them in your mind. As your ability to hold the pattern in your mind increases, you can flash the pattern, hold it in your mind, and then visualize what the pattern would look like from the three viewpoints (right, top, and left).

9

VISUAL FITNESS SKILL VII: SPORTS-SIGHT

Keeping Your Seeing on the Ball

Just as reading requires our seeing to be clearer, bigger, deeper, stronger, quicker, and smarter, all at the same time, so do sports. True, body strength, speed, agility, and coordination all contribute to success on the playing field or tennis court, but it's visual fitness that leads that strength, speed, agility, and coordination.

Because I'm a visual fitness coach, not a tennis or golf pro, I'm not going to supply you with any tips on your game. We'll leave that to the appropriate professionals. The goal here is to build your visual fitness so that when you do visit those pros you can fully profit from their help.

For peak sports performance, your technique must be perfected beyond conscious awareness. If you have to think about the components of your swing, you're not really being there. The time to perfect your technique is in practice. The same is true about the use of your eyes. The time to build your visual fitness is off the court or playing field so that when you do play, your seeing is second nature, something that just happens.

Before we begin building your Sports-Sight muscles, let's take a moment to review how your other visual-fitness muscles affect sports. Even if you have no interest in sports, what follows could as easily apply to general coordination, smoothness of movement, or confidence and safety when driving.

Clear-Sight

Baseball legend Ted Williams claimed he could see the seams on the baseball when he was batting. Considering that he was the last player in the past sixty years of major league ball to bat over .400, there's no reason to doubt him.

In sports, seeing the details increases our skill in pinpointing position, speed, and direction. Seeing the tip of the football is more likely to bring the ball into our hands than just seeing the ball. Seeing the seams of the tennis ball is more likely to help us contact the ball with the "sweet spot" of the racket. Obviously, we can't hit or track the seams we can't see. Having either perfect eyes or perfect contact lenses is essential for maximum performance in sports.

Notice that I said contact lenses, not glasses. There's a reason. Spectacle lenses introduce errors in seeing whenever we look away from the center of the lenses. If you're nearsighted—you can't see far away without glasses—and you move your eyes to look through the top of your spectacle lenses, an approaching ball will appear to be lower than it really is. If you look through the bottom of the lenses, then the ball will appear artificially high. The only way to receive correct information regarding position when wearing spectacle lenses is to look through the very center of the lenses. This requires head movement rather than eye movement, which is inefficient and disturbs balance.

As we learned in the Clear-Sight chapter, when it comes to how small you can see, perfect eyes are only the beginning. The player who can see letters half the size of the 20/20 line possesses twice the information about the ball. Not surprisingly, a military physical documented that Ted Williams had at least 20/10 vision the year after his .406 season. Despite this advantage, however, Williams knew there was more going on with his seeing than his eyesight. In his autobiography, *My Turn at Bat*, he wrote:

> Sure I think I had good eyesight, maybe exceptional eyesight, but not *superhuman* eyesight. A lot of people have 20-10 vision. The reason I saw things was that I was *intent* on seeing them. The reason I saw things was that I was so intense . . . I was intent on seeing them, I was looking all the time, I was alert for them. I trained myself . . . it was discipline, not super eyesight.

Intuitively, Williams knew that it wasn't his eyes, it was how he *used* those eyes that made for his success as one of the best, if not the best hitter in baseball history. Interestingly, he repeatedly used the terms *intent* and *intense* to describe the use of his eyes. If you want to see that same intensity in a modern-day athletic superstar, just look at the pictures in Tiger Woods's book, *How I Play Golf*. During Woods's address, his eyes are locked on the ball more intently than two heat-seeking missiles. Your Clear-Sight muscles control not only eyesight, but add to this intensity of seeing at any distance. If you want to optimize sports performance, your Clear-Sight exercises will help give you the seeing edge.

Big-Sight and Deep-Sight

Peripheral vision and depth perception are every bit as fundamental to sports performance as is Clear-Sight. In basketball, for example, Big-Sight is essential. Both Michael Jordan and Magic Johnson's sight was so big they seemed to have eyes in the backs of their heads. Big-Sight is important in other sports as well. In golf, for instance, reading a green requires not only the ability to see clearly enough to perceive the direction of the blades of grass, it requires seeing big enough and deep enough to take in the larger contours as well. The golfer who can integrate central and peripheral vision is at an immediate advantage. Big-Sight and Deep-Sight are, indeed, key components of the short game.

Similarly, in doubles tennis, Big-Sight allows us to instantaneously know the positions of our teammates and opponents without taking our eyes off the ball; Deep-Sight allows us to pinpoint the position of lobbed and overhand shots and to return them within the lines. Despite perfect form, the player with flabby Deep-Sight muscles will remain at a disadvantage during serves and overhand shots. Such players will have to hover at the baseline rather than face the increased visual challenges at the net. Problems with Big-Sight and Deep-Sight can handicap even the player who masters the technical skills.

Strong-Sight

Basketball Hall of Fame inductee Bill Russell once confided to author John McPhee: "The secret of shooting is concentration." Where concentration is concerned, Strong-Sight, no less than Clear-Sight, is indispensable.

The eye-muscle coordination of Strong-Sight comes in two varieties: coordination for near seeing and coordination for midrange and far seeing. Coordination for near seeing allows concentration during reading (providing you haven't learned to ignore one eye completely). Coordination for midrange and far seeing supports concentration in sports. The tennis player with flabby Strong-Sight muscles will be inconsistent and lose concentration in later sets. The golfer in need of Strong-Sight workouts will fatigue on later holes. When we are reading a book, our continued concentration requires not only Strong-Sight fitness but language skills. When it comes to sports, however, verbalization is more likely to get in the way. Strong-Sight is the key to maintaining concentration and performance, especially as the game progresses.

Quick-Sight

How quickly do truly great professional athletes see? A clue is provided by one of the finest basketball players of all time, Jerry West. According to author Andrew Cooper in *Playing in the Zone*, after West retired from playing and began coaching, he was frequently frustrated by the performance of his players. On one such occasion, he told Blazers center Bill Walton, "Many of the players rationalize their mistakes by claiming everything is happening so fast on the court that they get mixed up . . . but when I played, everything seemed to be happening in slow motion out there. I could see what was happening in advance, and anticipate plays."

Cooper found the same phenomenon discussed in Bill Russell's autobiography, *Second Wind*. Russell confided, "It was almost as if we were playing in slow motion . . . I could almost sense how the next play would develop and where the next shot would be taken."

Quick-Sight is no less important in baseball and tennis. In major league baseball, a pitched ball may approach the plate at one hundred miles per hour. In professional tennis, served balls may approach even faster. This phenomenal velocity leaves mere fractions of a second to see the ball and still have time to swing the bat or racket. The player who can see at a hundredth of a second what his opponents take a tenth of a second to see has a 1,000 percent advantage when it comes to hitting a baseball or returning a serve. In tennis, Quick-Sight added to Deep-Sight allows the player to be in the right place with enough time left to swing. The player who arrives in place too late to maintain good form while returning the ball is signaling the need for improved visual fitness.

Smart-Sight

As we learned in the last chapter, visualization is another of the keys to performance in sports. In *The Inner Game of Tennis,* W. Tim Gallwey describes a game in which "images are better than words," where seeing and visualizing replace damaging inner chatter. Similarly in *Visual Tennis,* John Yandell writes, "The simple fact is that tennis happens too fast to think about in words. But pictures can flow through the mind at the speed your body moves on the court."

According to Yandell, there are more than sixty studies that relate visualization to physical performance and surveys show that 85 to 95 percent of elite athletes use some form of visualization in their workouts. Before a game, for instance, tennis great Pancho Gonzales would sit in the locker room and visualize the moves he would need to beat his opponent. Superstar John McEnroe put it this way: "Sometimes in a match I'll suddenly see a shot flash across my mind's eye just before I hit it." This sounds much the same as Jerry West's and Bill Bradley's remarks about their ability to "see in advance" when in the heat of the game.

On the subject of visualization, Tiger Woods writes, "I believe that my creative mind is my greatest weapon. The best way to describe it is a kind of inner vision that enables me to see things others might not see." Similarly, Woods writes about a particularly challenging shot, "As with every shot I attempt, I visualized the ball's flight and how it should respond upon landing." About having made the shot, Woods continues, "Moments like that stay fresh in my mind, providing a positive image for future reference. Those images are critical when the game is on. They may even be the difference between success and failure."

Whether you're visualizing yourself as a winner or you're practicing your game in your mind, you want to keep your Smart-Sight muscles tuned up if you intend to play your best, no matter what the sport.

Sports-Sight

As you can see, all the visual fitness muscles we've discussed should be included in your workout if you're interested in sports. This said, what's left? What is it about sports that requires something more than what we've discussed?

Other than the distances involved, there is another major difference between sports and reading. When reading, both the reader and the book are presumably stationary. While this statement may be questionable in the case of some younger readers still blessed with an abundance of energy, or some older readers whose inappropriately small bifocals require head movement, for most of us, all that needs to be moving during reading is our eyes following the print and our hands turning the pages.

Sometimes this is true in sports, sometimes not. In golf, for instance, both the ball and the player's head are motionless, or nearly motionless, during the swing. In tennis, however, both the ball and the player are moving. The ability to maintain balance while keeping our eyes on the ball becomes essential. Similarly it becomes necessary to use our eyes to guide our hands and bodies. In ball sports, the player must rely on visual fitness to answer two questions simultaneously: Where is the ball? and Where am I? This second question includes, Where are my hands?

EYE-HAND COORDINATION

The term *hand-eye coordination* fails to describe the process of using our eyes and hands to catch a ball. Children go through a number of developmental stages in the use of hands with eyes. As infants, we seemed to see with our mouths; we gummed whatever found its way into our hands. Later we moved into hand-eye coordination. First, our hands chanced upon something; then, our eyes followed. As we matured, *eye-hand* coordination replaced the hand-eye stage; we spotted things with our eyes first and, only then, confirmed with our hands what we had seen. Finally, usually around the age of eight, the eyes took over and we became truly visual. From that point on, we could almost feel what we were looking at without using our hands.

Typically, it's not until we get into trouble with our seeing that we revert to an earlier level. The child who is going through the complexities of learning to read is likely to revert to using his finger for a guide. Such behavior is normal. The same holds true for adults. Flex your Smart-Sight muscles a bit and recall the last time you were in a clothing store and saw an unfamiliar fabric. Unless I'm mistaken, your first inclination was to reach out and touch the fabric. After that, no more touching was necessary; you could now feel that fabric with your eyes.

At any rate, visual fitness trainers frown at the term *hand-eye coor-*

dination because it suggests that the hands are guiding the eyes and while this may be true when a child uses his finger to learn to read, it's hardly true when we're catching a ball. If our hands are guiding our eyes, the ball is more likely to strike us in the chest than end up in our hands. In sports, it's our *eyes* that do the guiding. The football player who fumbles passes is said to have "bad hands," but the problem is with visual fitness, not hands. For this reason, visual fitness trainers speak of eye-hand, not hand-eye coordination.

BALANCE

To see how balance is related to visual fitness, try the following demonstration:

1. Balance on one foot. Be aware of how easy or difficult this is.

2. While maintaining your balance, close your eyes. Be aware of any change in your ability to balance.

Which was easier, balancing with eyes open or closed? The difference you experienced gives you some idea of the link between visual fitness and balance.

Our seeing and balance connect in another way as well. In the Big-Sight chapter, I asked you to perform a drill to open up your peripheral vision. Let's take a look at how big and little seeing relates to balance. Try this second demonstration:

1. Stand on one foot and make your sight big; be aware of the whole room in front of you and at the same time, as much of the wall, ceiling, and floor as you can see out of the corners of your eyes. How hard is it to maintain your balance this way?

2. While you continue to stand on one foot, look hard at something tiny in front of you: a detail on a painting or calendar, a screw in a switch plate. In other words, constrict your vision until it's as little as possible. What happens to your balance?

Repeat this demonstration a few times, trying to maintain your balance while shifting back and forth between big and little seeing.

If you were successful in shifting between your peripheral and central visual systems, you found that when you relaxed and were aware of the whole room your balance improved. When you constricted your awareness and looked hard at a tiny point, your balance grew more difficult to maintain.

Twenty percent of the nerve fibers that leave your eyes and travel into your brain do not contribute to what we normally think of as seeing. Instead, these fibers could be involved with coordinating our eyes, bodies, and balance. That you can manipulate this contribution by alternating your vision between big and little seeing shows further that not only eyes but how you use those eyes affects your balance. In other words, your balance is linked to your visual fitness.

KEEPING YOUR EYES ON THE BALL

So far we have considered both eye-hand coordination and balance as part of Sports-Sight. Now let's look at perhaps the most important component of all. Just as self-improvement gurus ask us to "visualize" but neglect to tell us how to form the pictures in our minds, so do coaches implore us to keep our eyes on the ball, but fail to provide the drills for improving this skill.

Saccadic Eye Movements

Players use several types of movements to keep their eyes on the ball. In the Deep-Sight exercises, for instance, I asked you to practice jumping your eyes from one corner of the room to the next. Eye doctors call this jumping type of eye movement a *saccade*—from a French word meaning "jerk." When our eyes make a saccade, they jerk from one side to the other. Saccades are voluntary eye movements. By this, we mean that you can perform a saccade whenever you wish. If you're looking at one corner of the room and you decide to look at another corner of the room you make a saccade—also called a saccadic eye movement.

Saccades have their advantages. For one thing, they're fast. For another, we suppress or ignore the swirl of the world during the instant we're making a saccadic eye movement. Not so when we move our heads. To experience this statement for yourself, leave your head stationary and jump your eyes back and forth between the first and last words of a single line of print on this page. Now repeat this movement, but instead of moving your eyes move your head. Notice how when you move your head rather than your eyes, the print seems to sweep back and forth in the opposite direction of your head movement. When you move only your eyes, the print appears stationary. This stable world is the result of your eyes blocking out the swirling images during the saccadic eye movements.

Readers who have to move their heads rather than their eyes—whether because of faulty eye movements, poorly designed bifocals, or too closely held books—are at a tremendous disadvantage. The same holds true for athletes. When Tiger Woods addresses the golf ball, his head is as still as the Sphinx's.

Pursuit Eye Movements

While we may use saccades to get our eyes on the ball in the first place, *keeping* our eyes on the ball requires other types of eye movements controlled from different areas of the brain. One such type of eye movement is called a *pursuit* because the eyes "pursue" or follow a moving target. Pursuits are smooth, not jerky—providing that the target being followed is moving smoothly and not too fast. The key here is that the eyes latch on to and follow the object. If there's no moving target to follow, there's no pursuit eye movement. If the object is moving too fast so that our eyes slip off it, we use saccades to get our eyes back on the object.

To experience this pursuit type of eye movement for yourself, extend one hand and arm away from your body. Close your fist, but leave your thumb pointing up in the air as if you're giving someone the thumb's-up. Keeping your arm straight, move your fist in a circle in the air at a constant distance from your face. While your hand revolves, keep your eyes on your thumb. So long as your eyes don't slip off the thumb, you're making a pursuit eye movement.

Now that you've experienced a pursuit eye movement for yourself, let's perform a couple more demonstrations that will increase your understanding of the role of such eye movements in sports.

After completing a good visual fitness program, athletes frequently comment that the ball doesn't seem to move as fast as it used to or that they have more time to see the ball. Part of this change comes from increased Quick-Sight fitness. Part of the change, however, is due to something else. To understand this "something," perform the next demonstration:

1. With your arm extended, move your thumb in a circle as you did above. Gradually increase the speed of the rotations while you keep your eyes locked onto your thumb. Be aware of how clear the thumb is and how fast it appears to be moving.

2. While you continue your arm rotations at the same speed, take your eyes off your thumb and stare straight ahead at some point across the room. Be aware of how fast the thumb now appears to be moving each time it sweeps across your line of sight. Be aware of how as the thumb passes, it's little more than a blur.

3. Again, without changing the speed of your thumb motion, lock your eyes back on the revolving thumb. Notice how the speed of the thumb seems to lessen as its details once more grow clear.

This is an excellent demonstration of why coaches tell us to keep our eyes on the ball. When we do, the speed of the ball seems to reduce and we have more time to catch or hit it.

Football coaches will sometimes ask receivers to zero in on the tip of the football rather than to just watch the ball. As mentioned before, Ted Williams looked for the seams on the ball; so do many tennis pros. This zeroing in on the details of the ball ensures that our eyes are, indeed, on the ball. To experience this for yourself, try the following demonstration:

1. Open this book to the chart on page 37, which you used to check your near eyesight in the Clear-Sight chapter. Use both hands to hold the book, so that the page remains open to the chart.

2. Begin to rotate the book slowly in a circle much as you did when rotating your thumb in the previous demonstration. Lock your eyes on the bottom line of the chart and attempt to keep that line clear.

3. Gradually turn the book faster, while you continue to watch its bottom line. Notice how if you take your eyes off the chart for even an instant the line blurs. It's only by keeping your eyes on the chart that you can continue to see the line clearly.

In this demonstration you are more or less measuring your *dynamic acuity* (*acuity* being the term for how small you can see, and *dynamic* referring to "movement"). During a regular eye-chart acuity test, the letters are stationary; during a *dynamic* test the letters are moving. While a stationary test may be fine to use for prescribing eyeglasses, a dynamic test is a far better measure of how we use our eyes, a better indicator of our visual fitness. Unfortunately, standardized dynamic acuity tests are not available for use in the doctor's office.

Vergence Eye Movements

Besides saccades and pursuits, there is a third type of eye movement. During our Strong-Sight workouts we practiced moving our eyes from far to near and near to far. We call these movements—in which our eyes converge toward our noses or diverge to point across the room—*vergence* eye movements.

Unlike saccades, which rely on voluntary effort, and pursuits, which rely on reflexes, vergence eye movements can involve either. If we decide to look from far to near, we can voluntarily converge our eyes. If we are tracking an object, such as a ball that is approaching us, then reflexes come into play that help keep both eyes aligned on the ball. Because both eyes have to be on the ball for depth perception to be fully realized, vergence eye movements are crucial for performance in sports. The tennis player serving, the golfer teeing off—both have to align their eyes precisely to avoid swinging long or short.

Those professionals who can see the seams of the ball, as Ted Williams once did, truly have their eyes on that ball, and as a result have all the time in the world to swing. Similarly, when Tiger Woods locks his eyes on the unmoving golf ball, he's using those eyes to guide one of the most effective swings in the world. Thus, the exercises below will help you learn to improve the coordination of your eyes, hands, and balance, not to mention your ability to keep your eyes on the ball.

Your Sports-Sight Workout

Sports-Sight I

This exercise works much the same as the Big-Sight exercises in that it forces us to integrate our central (intense) and peripheral (loose) vision. Sports-Sight I has seven steps:

1. Stand in a balanced position, your feet spread a little wider than your shoulders. Face a wall some feet in front of you as you did when working your Deep-Sight muscles. Gently cover your left eye with your cupped left hand. Extend your right arm until your closed hand—thumb up—is about sixteen to twenty inches in front of your face.

2. While you watch your thumbnail, move your right hand and entire right arm in a full circle, about a foot in diameter. Begin slowly. Work clockwise for a few revolutions, then counterclockwise for a few. As you move your hand, try to maintain your attention on your thumbnail. Be aware if at any time your attention or eye seems to wander from the thumb so that you have to reestablish or adjust your eye contact. If there is some particular direction where this lapse of concentration occurs—say, from the three to the five o'clock position—then work back and forth through this area a number of times until you can move through it without losing concentration.

3. As your ability to maintain eye contact and concentration (they're really the same thing) improves, open up your vision—without losing your concentration on your thumbnail—and become aware of where the wall meets the other two walls, the ceiling, and the floor.

4. After working for about two minutes with your right eye, switch. Cover your right eye with your right hand and rotate your left hand, following it as before, but this time with your left eye. Again, look for any lapses in concentration, any times you have to adjust your eye to get back on track. As your ability improves, open up your side vision as well and be aware of the wall behind your hand—without losing your concentration on your thumbnail. Take note of which eye is most difficult to use and in future workouts spend a little more time with the more difficult eye, a little less time with the easier-to-control eye.

5. After you have worked with your left eye for about two minutes, uncover both eyes. Rotate your preferred hand. Begin with your attention on the thumbnail, but gradually—without losing your concentration on the nail—open your attention to include an awareness of the wall in the background and the space between you and your thumb and between your thumb and the wall. Work this way for about two minutes. Try to open up your vision until you can follow your thumb with your eyes without the wall behind appearing to move. The more space you can see, the more stable the wall will be.

6. As your ability to maintain central concentration on your thumb improves while you're simultaneously being aware of the wall and space, alter your posture as follows: Instead of standing with your feet side to side, move your right foot forward until your two feet are aligned one in front of another, as if you were walking a tightrope. To maintain balance with your feet in this position, you'll have to really open up your peripheral vision and be aware of the space around you. It may be that aligning your feet one in front of the other is too difficult. If so, work on moving the forward foot into alignment with the rear foot gradually instead of all at once. For instance, the front foot might be allowed to rest a width or two to the side of the rear foot. If you are too central and lose your Big-Sight and Deep-Sight you'll lose your balance. With your feet in this position, you may have to move your circling hand very slowly, at least when you begin.

7. Repeat step 6, but this time with your left foot in front.

If any of the above steps are difficult, go back to an earlier step or an earlier part of a step until you have really mastered that before moving on. The above seven steps could take some weeks or even months to accomplish. Just remember to work where you can succeed. During each Sport-Sight workout, try to spend some time with each eye, some time with both.

Sports-Sight II

This exercise is much like the last, but with one major exception: Instead of following your thumb, you're going to follow the Clear-Sight Near Chart A from the Clear-Sight chapter (on page 47).

1. Begin as you did in steps 1 and 2 of Sports-Sight I, but instead of holding your thumb out in front of you hold Near Chart A.

2. As you rotate the chart, read the lines. Begin slowly. Be aware if your concentration ever breaks, making you lose your place on the chart or the letters to become blurred or your eye to fall off the chart, and you have to move your eye back to the letters. If you reduce the diameter of your rotation a bit, you can do this exercise with regular reading glasses. If you're not too nearsighted, you can move the chart a little closer and do the exercise without glasses. If you're wearing bifocals, however, it is unlikely that you will have room to make a large enough rotation for this drill to work. If this is the case, you may need to take off your glasses and use the Clear-Sight Far Chart (page 45) to have letters large enough to see.

3. Work Near Chart A with your right eye, left eye, and both eyes. First work on accuracy, then on speed, then on opening up your side vision without losing accuracy or speed.

4. Work until you can keep the chart clear with both eyes open while you move the chart rapidly. Also maintain awareness of the space (depth) between yourself and the wall and yourself and the chart. As your skill improves, gradually rearrange your foot position until one foot is directly in front of the other.

5. When you have mastered the first four steps, repeat them again, this time with Clear-Sight Near Chart B (page 48).

6. Finally, try this variation: Instead of reading the letters in the lines of the charts straight across, read the first letter, the last letter, the second-to-the-first letter, the second-to-the-last letter, and so forth. For example, in the first line you would read, "E, P, F, D, T, V, Z, C, A." You'd start with the outside letters of the line and work your way inward. After you completed the first line, you'd continue down to complete the chart. With the procedure modified in this manner, perform the following two steps:

A. Practice your new side-to-side method of reading the chart without rotating it. Time yourself and continue working until you can read down the entire chart in less than fifty seconds.
B. Once you have mastered the side-to-side reading of the chart, begin to rotate the chart. Work through steps 1–4 of Sports-Sight

II, until you can read the chart in a side-to-side fashion while seeing big and standing with one foot in front of the other.

By the time you learn to combine speed, balance, clear central vision, and awareness of depth and periphery, as well as mastering the six other visual fitness skills, you should be ready for anything a coach—or highway—sends in your direction.

10

Your Visual Fitness Workouts: Organizing for Success

The number-one secret to success in your visual fitness program is to do the exercises. Do them, and the results will come. Reading the book may allow you to understand the source of many frustrations you've faced in life, but only by doing the exercises will you expand your visual fitness.

The best way to get the exercises done is to schedule them in advance. Time is like money. Those who manage to save put money aside at the beginning of the month. Those who wait until the end of the month to see what's left usually save little. Saving time for your fitness program is much the same. If you save your workouts until you have time, believe me, there will never be enough time. Therefore, to make this program work, you have to save your time in advance and plan your workouts. Try scheduling your workouts for every day before breakfast, or Monday, Tuesday, Thursday before dinner. If you're a morning person, schedule them for before the sun comes up. If you're a night owl, schedule them for later in the evening when you get your second wind. Consistency is the key.

How much time?

That depends on how quickly you want to see results. The typical program takes about thirty hours, although some may need to spend twice that much time, depending on their initial level of visual fitness.

As I'll show you later, there are any number of ways you can divide up those hours. It all depends on how much your visual fitness is getting in the way and how quickly you want things to improve. The least involved method will take just seven minutes per day.

But What if There's No Time?

If you lack the time to do your exercises, go back to chapter 2. Look over your Visual Fitness Self-Assessment and ask yourself the following questions:

- How much time am I throwing away because of poor visual fitness?
- How much time is lost because I'm exhausted?
- How much time is lost to sore eyes or headaches?
- How much time is lost at work?
- How much time is lost in my relationships because I'm irritable?
- How much time is lost in my reading because I can't stick with it?
- How much time is lost misjudging the position of the ball in sports?
- How much time is lost because I have to look at everything twice just to make sure?
- How much time could I lose if I had a serious automobile accident?

Add up the total time you are losing or could lose. If that total exceeds seven minutes a day, then you'll actually save time by doing your exercises.

Will It Work for Me?

A good question. Frequently when people are concerned about the time commitment to improve visual fitness, they are actually concerned with the possibility of wasting their time. In other words, they're worried that they'll spend the time and not get anything for their effort.

In our office, visual fitness training has worked for thousands. The studies in Appendix B suggest this is no fluke. Whether or not the program will work for you, however, is a different question. As we mentioned repeatedly in chapter 1, this book is designed to help those who already have perfect eyes or perfect glasses. If you have some sort of eye disease, this fitness program is not your answer. If you're three years late for your yearly eye exam and your glasses are beginning to look as worn as the pyramids, maybe you should drop by and see the doctor. If you have an eye that drifts in toward your nose or out toward your ear, you're going to need a professional trainer. If you're long on desire but short on self-discipline, you're similarly going to need a trainer. If you want to throw away your glasses, I'll have a better book for you in the future. But if you already have perfect eyes or perfect glasses and you want your seeing to be clearer, bigger, deeper, stronger, quicker, smarter, then your success with visual fitness will depend largely on your attitude—that is, on how you see the world.

How about it? How do you see the world? Is it half full and on its way to being fuller, or is it—like the proverbial glass—half empty? If, after reading the previous chapters, you now see your world as on its way to being fuller, you're going to do fine; you'll be able to see all the little improvements that add up to success. If however, your world looks half empty, and getting emptier all the time, then you're likely to have a harder time of it—at least until you relearn how to see. But then that's one of the beauties of visual fitness: when we question our visual athletes about the changes they've seen since beginning their workouts, "improved self-confidence" and "more positive attitude" are two of their most common responses. And no wonder: improving visual fitness changes the way we see the world. Even the half-empty visual athlete has every chance for a fill-up.

But then, I guess I'm preaching to the choir. That you picked up this book in the first place suggests that you're already an on-the-way-to-being-fuller person and that the exercises in the book will help you in your quest to fill your life to the brim.

Seven Minutes a Day

The seven-minute-a-day program is designed for those who feel they're doing well already but could be doing better. If you fall in this category, just schedule a time of day to do your exercises and get started.

Your first time through the workouts, spend three days on each of the seven visual fitness skills including Clear-Sight, Big-Sight, Deep-Sight, Strong-Sight, Quick-Sight, Smart-Sight, and Sports-Sight. When the three days of working Clear-Sight are over, move on to Big-Sight even if you haven't yet mastered all the Clear-Sight steps. After three days on Big-Sight, move on to Deep-Sight, and so forth. Continue this schedule until, at the end of three weeks, you've rotated through all seven skills. Then begin again. If during your rotations you'd prefer a day of rest, then take off one day a week and double up on some other day so you can continue to get in your seven workouts per week.

Each time you pick up a formerly worked skill, spend your first day working on the last step you had mastered on your previous rotation. For instance, there are six Big-Sight charts. If on your first three days you had mastered four charts and had begun working Chart V, then, when you resume your Big-Sight workout three weeks later, begin with Chart IV. If ever you find yourself struggling with a step, drop back to an earlier step or exercise.

Continue the three-week rotations, working each visual fitness skill three days at a time until you begin to master the skills, then modify the sequence in the following manner: The next time you rotate onto the mastered skill, spend only one day. For example, suppose you master all six Big-Sight charts to your satisfaction. The next time you rotate onto this skill, you will spend only one day on Big-Sight VI before moving on to Deep-Sight. You will continue this sequence until you have mastered all seven categories. At this point, you'll be rotating through the seven skills weekly.

Accelerated Schedules

Seven minutes a day will get you there, eventually. If, however, you believe time's a-wasting, you may want to accelerate your progress. Say for instance that you're sick to death of the daily headaches, or the fatigue when reading, or the urge to snap at your loved ones when you get home from work. Or say that you're worried about driving at night, or frustrated that for what you've spent on tennis lessons you could own your own court, yet you're still contacting more balls with your forehead than your racket. Whatever your reason, you can speed your visual fitness progress along by extending the length of your workouts. For example, you could include all seven skills in the same workout. If you allowed a minute or two between skills, such a workout would take about an hour.

To double your progress compared to the seven-minute schedule, you'd simply perform these hour-long workouts twice a week. To triple your progress, you'd exercise three times. So long as you don't spend more than seven minutes per workout on any single skill—except for maybe Quick-Sight or Smart Sight—you can work out as many days per week as you wish. As for the days: Choose any days that work. Monday, Wednesday, Friday would be a sample schedule, but Monday, Tuesday, Wednesday could work equally well. You don't need to space the days. In fact, if Saturday were the only day available, you could do two one-hour workouts on the same afternoon. Just be sure to take an hour break between workouts.

If an hour workout is inconvenient, you could still double your progress by working thirty minutes a day, four days per week, or even by doing two seven-minute workouts a day. Instead of working Clear-Sight twice, work Clear-Sight and Big-Sight for three days then move forward to Deep-Sight and Strong Sight. Again, the key to success is to schedule the workouts in advance and to make an agreement with yourself to follow the schedule.

In addition to your scheduled workouts, you can perform Deep-Sight III for a few minutes whenever you're moving toward a visual headache.

Your Workout Record

To help you organize your time and keep track of your progress, use the "Workout Record" provided on page 199. The seven visual fitness skills are listed down the left side.

Each time you complete seven minutes on a particular skill, record the date and the level reached in one of the squares next to that skill. For example, if you're so determined to gain your visual independence that the Fourth of July finds you working step 6 of Strong-Sight I, you would write *July 4: I-6* in the box next to Strong-Sight.

If you're doing the seven-minute-per-day program, you will fill up the Clear-Sight row, then move on to the Big-Sight row. If you're following an accelerated three one-hour workout program, you will fill out the seven boxes in the first column all on the same day, as your workout progresses. On the next workout, you'll fill out the seven boxes in the second column, and so on.

Okay. So much for the basics. Get started with Clear-Sight I. If you're doing the accelerated program, it may be best to start your workouts with two or three visual fitness skills, then gradually add the others as you learn them until you're working all seven skills during the same workout.

Good luck.

Workout Record

SKILL I **Clear-Sight**			
SKILL II **Big-Sight**			
SKILL III **Deep-Sight**			
SKILL IV **Strong-Sight**			
SKILL V **Quick-Sight**			
SKILL VI **Smart-Sight**			
SKILL VII **Sports-Sight**			

11

Visual Fitness Trainers: Who Are They? Where Did They Come From? Why Should You Visit One?

Is visual fitness training new?

Compared to medicine, yes. Physicians began treating crossed eyes medically almost four thousand years ago. As M. J. Revell discusses in his history on eye exercises, an ancient Egyptian record called the *Ebers Papyrus* (written at the time of Moses) suggests, "For the turning of the eyes equal parts tortoise brain and oriental spices rubbed together." It was not until 1896, four millennia later, that French visual scientist Emile Javal—whose cross-eyed father had been disfigured by eye surgery and whose sister had inherited the same condition—published the first book on using exercises to treat crossed eyes. Across the English Channel, British eye surgeon Claude Worth published a similar book in 1903.

As the twentieth century progressed and surgical technique improved, medical eye doctors—ophthalmologists—continued to argue within their own ranks about the relative merits of eye exercises versus eye surgery. In the 1950s, it was common for a cross-eyed child to begin with eye exercises, have a single surgery, and then do more eye exercises. By the 1960s, however, two major changes had taken place: anesthesia had advanced to make surgery safer, and major medical insurance coverage had arrived to make surgery affordable. From that period on, ophthalmologists gradually abandoned the combined exercise-and-surgery approach and moved to a more convenient and less time-consuming multiple surgery approach.

Today, some of the more progressive and open-minded ophthalmologists are reconsidering the possible role of eye exercises in the treatment of eye turns and even visually related reading problems. Medical doctors remain cautious, but over the past ten years, several articles (see Appendix B and the Bibliography) have appeared in the medical journal *Binocular Vision and Strabismus Quarterly* ("binocular" meaning two-eyed, and "strabismus" meaning eyes that are not aligned) edited by Dr. Paul Romano. For the most part, however, if patients want the alternative of eye exercises, they still have to look outside of medicine.

Optometry is the eye-care profession that is independent of medicine. Just as dentists are trained in a system separate from but parallel to medicine, so are optometrists. Today, in all fifty states, optometrists not only prescribe glasses and contact lenses, but use medications to treat eye disease, much as dentists use medications to treat oral disease. In Oklahoma, optometrists even perform laser surgery. In the first half of the twentieth century, however, little of this was the case.

During that period, when ophthalmologists were honing their surgical techniques, optometrists had neither drugs nor surgery at their disposal. Thus, they spent their time in other directions. One of those directions included building upon and improving the earlier eye-exercise successes of Javal and Worth.

At the beginning of the century, the exercises were limited to what I have called the Strong-Sight muscles, or what eye doctors refer to as *binocular* or two-eyed seeing. These early exercises relied on instruments that excluded peripheral vision; the Big-Sight skills were ignored, as were all of the other visual fitness skills we covered in chapters 3 through 9. But all of that was about to change.

In the 1920s a young optometrist named A. M. Skeffington listened to his colleagues' theories on the measurement of eyes and realized that something more was going on. In his own practice, Skeffington was observing that all too often people were visually inefficient even though the glasses he prescribed them were consistent with their eye findings. Intuitively, he sensed that vision wasn't just a passive process related to the measurements of eyes. He knew there was more involved. In 1928, Skeffington joined another optometrist, E. B. Alexander, to form the Optometric Extension Program (OEP). The organization specialized in research and education in vision. Soon the OEP had three thousand members and until the 1960s was the primary force in postgraduate optometric education.

Some of the best minds in the profession researched eye exercises in the optometry school clinics and laboratories. Others joined in Skeffington's search to understand not just eyes, but also how we use them. These optometrists interacted with psychologists, educators, physiologists, and physicians in an effort to understand what they were seeing with their patients. As their understanding of how we take in information through our eyes increased, so did the scope of their eye exercises.

To them, *vision* became the word used to describe what they considered to be the dominant sense, which steered the others. According to Skeffington and his associates, *vision* didn't exist in the eyes, it was a process that involved the total human organism. *Vision* wasn't fixed, but constantly changing and developing. *Vision* could be warped when stress was added to seeing and could eventually contribute to eye and eyesight problems. *Vision* was active, not passive; something we did, not just something we experienced. *Vision* was learned, and therefore it could be taught. *Vision* wasn't merely eyesight, it was everything and more that I've described in this instruction manual for using our eyes.

What Do Optometrists Who Train Visual Fitness Call Themselves?

While there are many optometrists who offer eye exercises along the more classical Strong-Sight–oriented models of Javel and Worth, the optometrists who subscribe to Skeffington's expanded definition of *vision* call themselves behavioral optometrists. According to the late Dr. Martin Birnbaum, professor in my residency program at the State University of New York, the term *behavioral* rests on the model that "our *behavior* influences how we see, and how we see influences our *behavior.*"

The way in which behavioral optometrists view vision has vastly expanded their approach to treating and enhancing the way we use our eyes. Their use of the word *vision* to include more than eyesight, however, creates a problem: it confuses just about everyone. To the rest of the world *vision* means calling out those itsy-bitsy, teensy-weensy little letters on an eye chart conceived almost one hundred and fifty years ago. Even general optometrists and ophthalmologists use the terms *vision, eyesight,* and *acuity* interchangeably. The moment we hear the word *vision,* we think about clarity of seeing, not about our seeing being clearer, bigger, deeper, stronger, quicker, and smarter. The moment we hear the word *vision,* we immediately assume, if our eyes or glasses are perfect, that the topic concerns us no more than would the history of boxer shorts on polar bears in South Georgia. The moment we hear the word *vision*, we close our eyes—and minds—to our potential for using those eyes to interact with the world.

The same is true when we hear of *vision* exercises: we think of clear seeing and assume the exercises have something to do with eyesight, possibly some alternative to eyeglasses like that proposed by Dr. William Bates in his 1940 book, *Better Eyesight Without Glasses*. While optometrists have helped thousands to improve eyesight without glasses and have advanced the approach considerably since Bates's time, optometric eye exercises offer far more than enhancing clear seeing.

For all of these reasons, I've substituted the term *visual fitness* for *vision* throughout this book.

What Is the Technical Term for Visual Fitness Training?

In the early 1900s when ophthalmologists first used eye exercises to straighten crossed eyes, they called the process orthoptics, from the Greek *ortho*, meaning "straight," and *optikos,* which is related to seeing and the eyes. Alone, the term *orthoptics,* or "straight eyes," was descriptive enough initially, but when optometrists began to do more than straighten eyes, they came to prefer the term *vision training*, with *vision* carrying its expanded definition and significance. In time, however, the optometrists supervising the exercises realized that they were training more than vision, they were expanding human potential and performance. Therefore, they came to prefer the term *visual training*, meaning that the training they were using to expand lives was *visual* in nature.

Since the 1970s, optometry has moved more in the direction of medical thinking, which includes noting symptoms, finding a diagnosis that explains the symptoms, and then treating the disease or disorder diagnosed. To accompany this shift in thinking away from expanding human potential and toward that of treating disease, most behavioral optometrists now use the term *optometric vision therapy.* According to the third edition of *Webster's New World College Dictionary*, the term *therapy* means "the treatment of disease or of any physical or mental disorder." If you have an actual eye-muscle problem, such as strabismus ("crossed" or "wall" eyes) or convergence insufficiency or convergence excess (you can't aim your eyes accurately at closer distances), the term *vision therapy* certainly fits. And, as I've mentioned repeatedly, treating eye-muscle problems is beyond the scope of the visual fitness enhancement procedures described in this book.

If Optometric Vision Therapy Has Been Around for Almost a Century Why Haven't We Heard More about It?

The answer is quite simple, really. While hundreds of thousands of people have benefited from vision therapy, the doctors who provide the service haven't spent enough time and money promoting what they do—and there is no related industry that promotes the service for them. Toothpaste companies spend millions telling us to see the dentist. Pharmaceutical companies spend even more telling us to see our physicians. In eye care, Johnson and Johnson and Ciba Vision make sure we've heard of the latest in tinted and disposable contact lenses. And when it comes to refractive surgery, a single laser goes for a quarter of a million dollars, so there's plenty of money spent on telling us how we can now throw away our glasses.

Until there are more optometrists providing vision therapy, or until we start to find visual fitness exercises on the back of cereal boxes, the dollars just aren't available to educate the public fully on the benefits of vision training.

Why Do So Few Doctors Provide Vision Therapy?

Few optometrists provide comprehensive vision therapy services, probably for the same reason that few Ph.D.s go on to become coaches. Interests vary, and motivating others to change themselves isn't for everyone. In addition, most vision therapy patients are children, which requires even more energy.

A story that eye-exercise historian M. J. Revell tells concerning Javel, the Frenchman who wrote the 1896 book on straightening eyes without surgery, provides a good example of the amount of energy involved. Javel developed his exercises to help his cross-eyed sister avoid the primitive surgeries of the day, what he termed *le massacre des muscles oculor-motor,* loosely translated: "the massacre of the muscles that move the eyes." Javal's exercises, however, were exceedingly time consuming and Albrecht Von Graefe, a noted surgeon of the period, said of the approach, "People are not worth the effort." Half a lifetime later, at the end of Javel's career, he was going blind from excessive pressure inside his eyes. Feeling old and discouraged, he concluded, "Von Graefe was correct."

Unfortunately, the curriculum for modern-day optometry students is so packed, that many of them begin to feel as old as Javal before they finish their first year in professional school. By the time they graduate they are too busy getting started to devote the energy it would take to add vision therapy to their practices.

Fortunately, a percentage of students do gravitate to vision therapy. In my case, when I was an undergraduate at U.C.L.A. I majored in English literature at the same time that I took calculus, physics, and chemistry, the biological and behavioral sciences that formed my pre-optometry requirements. In my initial year of optometry school one of the first things I was taught was that when it comes to examining eyes, twenty feet is the same as infinity. It makes no difference whether we view the gleaming of the infinite or the eye chart at the end of a twenty-foot room; either way what I have called the Clear-Sight muscles in this book completely relax in the same manner.

By the time I had finished my first year surrounded by those who now accepted that infinity was twenty feet, I felt a little claustrophobic. I was lukewarm about the prospect of a lifetime spent in a semidark room asking the quintessential optometric question, "Which is better: one, or two?" Perhaps I had trouble remembering the answer; I don't know. But then I met Dr. William Ludlam, a professor who illustrated his lectures with stories about children and adults whose lives had been changed as a result of vision therapy. I was, at once, sold on the subject. It seemed to me that the art of vision therapy was the closest thing to the humanities in the increasingly scientific field of optometry. I'd found my niche.

Over the next year and a half, I donated hundreds of hours working in the vision therapy clinic. My first patient was a nine-year-old student whom I'll call Joe. The boy had an eye that turned in almost to his nose. He was legally blind in that eye, meaning that even when wearing the best pair of glasses his vision was no better than 20/200. Even worse, his brain had learned to process the information from his turned eye in an odd way that most experts of the day claimed couldn't be helped. Bill Ludlam disagreed.

For many months, Bill guided my work with Joe until the boy's eyes looked straight. Joe's sight in his left eye was vastly improved but still 20/30, two lines short of the sacred 20/20. In a strict sense, Joe had not been "cured."

When I informed his mother about the failure, she looked at me as if I were crazy and said, "What do you mean you failed? Joe can finally ride a bike; we can read his handwriting. He's no longer failing in school, and now in sports he isn't always left sitting on the bench. How could you call that a failure?"

How could I?

From that point on, I had a picture in *my* mind of Joe that saw me through all the difficulties of developing a vision therapy practice. Since then, I've seen thousands and thousands of children and adults who—even though their eyes were completely straight before they began—nonetheless experienced life-changing improvements in learning, earning, and sports, not to mention eased family tensions.

How Are Visual Fitness Coaches/Behavioral Optometrists Trained and Certified?

After college, optometrists attend four years of professional school. During those four years they receive course work and clinical experience in what they'll need to know to diagnose and treat eye conditions. Included in this course work and experience is some basic work in vision therapy. Optometrists then have to pass extensive National Board Examinations that cover all areas of nonsurgical eye care; the tests include basic questions on vision therapy as well.

Optometrists who graduate from professional school earn an O.D. degree, standing for Doctor of Optometry. Once licensed to practice, any optometrist can offer vision therapy. Most optometrists interested in this area of care, however, go on to take hundreds of hours of postgraduate education. A few even attend formal residencies, as I did.

Once these additional educational requirements are passed, interested doctors can complete a certification process offered by the College of Optometrists in Vision Development. The process requires case reports, written and oral examinations. Those who complete this process become Fellows of the College of Optometrists in Vision Development and now sport the additional initials F.C.O.V.D. behind their names. There is also a separate examination offered by the American Academy of Optometry. Once the doctor has passed this exam, he or she becomes a Diplomate in Binocular Vision and Perception. (*Binocular vision and perception* are technical terms that loosely describe "visual fitness," but, really, would you have picked up a book called *Binocular Vision and Perception: 7 Minutes to Better Eyesight and Beyond*?)

Are There Studies That Support Vision Therapy?

Yes. There are hundreds of studies that support vision therapy. You'll find a sample of fifty of them listed in Appendix B, "Research on Vision Therapy."

How Does Optometric Vision Therapy Differ from the Procedures in This Book?

Just as you'll learn to play tennis more quickly if you have the help of a pro, you'll learn to see more quickly if you have the help of a doctor. While the procedures in this book could all be considered examples of optometric vision therapy, the therapy you'd receive in a doctor's office would vary considerably from that presented here. For one thing, a doctor would tailor your program, selecting from hundreds of procedures rather than from the handful described in this book. Many of those procedures would make use of lenses and prisms, which alter the light entering your eyes in such a way that your old, bad seeing habits will no longer work. As a result, you'd be forced to discard those old habits more quickly and your progress would be more rapid. The doctor would also use a number of instruments that give you exact feedback when your eyes are working and when they are not. In the doctor's office, you would be guided every step of the way. The exercises in the book provide a mere fraction of what is available.

Therefore, if you feel that problems with visual fitness are significantly interfering with your goals in life, or if you have a child whom you believe may be struggling unnecessarily in school or sports or both, I'd suggest you contact a behavioral optometrist. You'll find information on how to find one in Appendix A.

12

Breaking the Sight Barrier: Seeing with the Power of Choice

Since the advent of the Internet, the world is expanding faster than we can see to keep pace, faster than the speed of sight. There's more reading to do than ever before. The most common reason that individuals come to see me is to break the sight barrier to comfortable and efficient reading. As we've discussed, other common goals include making night driving safer, being less fatigued at the end of the day, improving golf or tennis.

In addition, some see me to handle more serious vision problems including aligning crossed eyes or seeing better out of a lazy eye. Few patients arrive, however, asking, "Can you change my outlook on life?" And yet again and again, that's exactly what happens. This unlooked-for bonus accompanies the attainment of other goals.

It comes as no surprise. "Outlook" requires us to look outward, and that's exactly what visual fitness is all about. See for yourself; next time your outlook is poor, be aware of how difficult it is to focus your attention outward at the world around you. It's as if you're stuck inside your head. The solution can be as easy as a good visual fitness workout. Force your attention outward. Rev up your seeing until it's clearer, bigger, deeper, stronger, quicker, and smarter. Visualize yourself achieving some goal and toss out any pictures of failure; look out and see how much air fills the room around you. Then see if you don't feel better. You'll find that the more power you have over your visual fitness, the more choice you'll have over your outlook. The following story is a good example.

Seeing to Soar

John was a pilot in his late thirties when his airline closed its doors. Not only was he out of work, but, because of that period's 20/20 hiring policies, he was too dependent on glasses to find work with another major airline. John felt he was a victim of circumstances, that his airline and eyes had sealed his fate without allowing him any choice in the matter.

When he arrived at our office, a fatalistic philosophy seemed to have followed him. His stooped shoulders subtracted inches from his height of six feet. If you looked into his eyes, it was almost as if he were lost three feet behind his eyeballs; he was definitely stuck looking inward. His face was blank, almost solid looking as if it no longer had the flexibility left for a smile. His attitude was as inflexible as his face. He was skeptical that anything could be done for his eyes, and when he reluctantly signed up for a program, one had the feeling that he had done so just to prove that life had indeed singled him out for failure.

I wasn't so sure that he was wrong.

For a time, John made little progress. The exercises seemed beyond him. He wasn't surprised; he wasn't disappointed. He seemed resigned to his fate.

We continued.

We worked with John using some of the exercises included in this book and many additional techniques that required special lenses and instruments.

One of the instruments gave John feedback when he shut off, or suppressed, the information coming from his right eye. The target he watched contained an *R* that was seen by his right eye and an *L* that was seen by his left. For many sessions, John watched the *R* blinking on and off saying, "It went away. It came back." That the blinking of the letter had anything to do with him escaped John entirely.

Then one day something changed. I directed John's attention and asked him to look more out of the right side of his head—much as I asked you to do, when you begin Strong-Sight II. When John tried what I had asked, he saw the *R* momentarily appear and said, "I got it on!" For the first time, it wasn't just luck, it wasn't just chance. He realized that *he* had something to do with how he was seeing. When the *R* disappeared, however, his responsibility for his vision quickly vanished and he said, "It went away."

I was nonetheless pleased. John was halfway there. He was willing to take credit for the good in his seeing. The bad, however, was still attributed to something beyond him.

We continued working. Finally, when I directed John to look out of the left side of his head, the *R* disappeared and he said, "I made it go away!"

Before long, John could control the presence of the *R* at will. He knew he was causing the letter to appear and disappear. His seeing had begun to fall within his power of choice. From this point on, John made rapid progress. Soon he could choose to make his sight clearer, bigger, deeper, stronger, quicker, and smarter at will.

The more success John experienced with his visual fitness and the more his power of choice over his seeing improved, the less solid his face looked and the brighter his eyes became. Soon he was able to shift from looking inward to looking outward.

As John's outlook improved, he became more and more certain of his success; his progress continued to accelerate. Without waiting for his acuity to reach 20/20, he began applying to other airlines, confident that when the time came he'd see well enough to pass his physicals. With his new attitude, there was little doubt that he'd sail through his interviews.

Before the year had passed, John was once again flying. Just as he had regained his power of choice over the use of his eyes, so he had regained his power of choice over his occupation and life.

John is not alone. Over the years, I've seen thousands of patients whose outlook has improved along with their visual fitness. Countless men, women, and children have entered their programs convinced that they were stupid, only to see their reading improve and gain the insight that it was the use of their eyes, not their intelligence, that had been the problem all along.

As a visual fitness coach, I see this link between outlook and eye control too often to consider it coincidence. It's almost as if when patients learn they are responsible for the use of their eyes, they begin taking more responsibility for their lives as well. Of the 838 patients I surveyed for my 1995 study "Vision Therapy and Quality of Life," well over half mentioned such behavioral changes as improved "confidence," "self-confidence," "positive attitude," and "self-esteem"; improved "concentration" and "attention span"; reduced "fatigue," "frustration," and "anger"; and even "better eye contact" and "improved family relations." I didn't ask patients if there had been a change in any of these areas, I just asked, "What changes have you seen since beginning vision therapy?"

Not surprisingly, it works the other way as well. Those who already have positive attitudes at the outset begin seeing changes in how they use their eyes almost immediately.

If you're one of these positive people, and you now believe that better visual fitness would allow you to achieve some goal, then either get started on the exercises in this book or find a behavioral optometrist/visual fitness coach. Put in the time. Put in the effort, and you'll be well on your way to breaking the sight barrier to learning, earning, and fun.

I've enjoyed spending these pages with you. Good luck with your visual fitness, and if you have questions about your visual fitness or about finding a trainer, use the information in Appendix A to contact our office.

APPENDIX A

Finding a Visual Fitness Coach

If you'd like help finding a doctor who offers vision therapy, please visit my website or contact my office by mail, phone, or e-mail:

David L. Cook, O.D., F.C.O.V.D., F.A.A.O.
1395 S. Marietta Parkway
Building 400, Suite 116
Marietta, GA 30067
(770) 419-0400 or toll free (866) 268-3937
www.cookvisiontherapy.com
Cook2020@aol.com

Any licensed optometrist can provide vision therapy or can refer you to another doctor with an interest in vision therapy. In addition, there are two organizations that provide advanced testing in vision therapy for doctors who have already graduated from optometry school. Both organizations provide a directory of doctors who have passed such postgraduate testing:

College of Optometrists in Vision Development
243 North Lindbergh Boulevard, Suite 310
St. Louis, MO 63141
(314) 991-4007 or toll free (888) 268-3770
www.covd.org

American Academy of Optometry
6110 Executive Boulevard, Suite 506
Rockville, MD 20852
www.aaopt.org (Click on "Member Directory." Look under "Diplomates." Search under "Binocular Vision and Perception and Pediatrics.")

You'll find the best source of books and other written material on vision therapy through:

Optometric Extension Program Foundation (OEPF)
1921 East Carnegie Avenue, Suite 3L
Santa Ana, CA 92705
(949) 250-8070
www.oep.org (Click on "Resource Center")

For more information on doctors providing optometric vision therapy, you can also contact:

American Optometric Association
243 North Lindbergh Boulevard
St. Louis, MO 63141
(314) 991-4100
www.aoa.org

APPENDIX B

Research on Vision Therapy

Atzmon, D., P. Nemet, A. Ishay, and E. Karni. A randomized prospective masked and matched comparative study of orthoptic treatment versus conventional reading tutoring treatment for reading disabilities in 62 children. *Binocular Vision and Eye Muscle Surgery Quarterly* 8 (1993): 91–106.

Birnbaum, M. H., K. Koslowe, and R. Sanet. Success in amblyopia therapy as a function of age: A literature review. *American Journal of Optometry and Physiological Optics* 54 (1977): 269–75.

Birnbaum, M. H., R. Soden, and A. H. Cohen. Efficacy of vision therapy for convergence insufficiency in an adult male population. *Journal of the American Optometric Association* 70 (1999): 225–32.

Buzzelli, A. R. Stereopsis, accommodative and vergence facility: Do they relate to dyslexia? *Optometry and Visual Science* 68 (1991): 842–46.

Ciufredda, K. J., S. G. Goldrich, and C. Neary. Use of eye movement auditory feedback in the control of nystagmus. *American Journal of Optometry and Physiological Optics* 59 (1982): 396–409.

Cohen, A. H., and R. Soden. Effectiveness of visual therapy for convergence insufficiencies for an adult population. *Journal of the American Optometric Association* 55 (1984): 491–94.

Cooper, J., et al. Reduction of asthenopia in patients with convergence insufficiency after fusional vergence training. *American Journal of Optometry and Physiological Optics* 60 (1983): 982–89.

Cooper, J., and J. Feldman. Operant conditioning of fusional convergence ranges using random dot stereograms. *American Journal of Optometry and Physiological Optics* 57 (1980): 205–13.

Cooper, J., and N. Medow. Intermittent exotropia basic and divergence excess type. *Binocular Vision and Eye Muscle Surgery Quarterly* 8 (1993): 185–216.

Cooper, J., and R. H. Duckman. Convergence insufficiency: Incidence, diagnosis and treatment. *Journal of the American Optometric Association* 49 (1978): 673–80.

Cornsweet, T. N. Training the visual accommodative system. *Vision Research* 13 (1973): 713–15.

Daum, K. M. The course and effect of visual training on the vergence system. *American Journal of Optometry and Physiological Optics* 59 (1982): 223–27.

———. A comparison of the results of tonic and phasic training on the vergence system. *American Journal of Optometry and Physiological Optics* 60 (1983): 769–75.

———. Predicting results in the orthoptic treatment of accommodative dysfunction. *American Journal of Optometry and Physiological Optics* 61 (1984): 184–89.

Duckman, R. H. Effectiveness of visual training on a population of cerebral palsied children. *Journal of the American Optometric Association* 51 (1980): 1013–16.

Etting, G. Strabismus therapy in private practice: Cure rates after three months of therapy. *Journal of the American Optometric Association* 49 (1978): 1367–73.

Farrar, R., M. Call, and W. C. Maples. A comparison of the visual symptoms between ADD/ADHD and normal children. *Optometry* 72 (2001): 441–51.

Flax, N., and R. H. Duckman. Orthoptic treatment of strabismus. *Journal of the American Optometric Association* 49 (1978): 1353–61.

Gallaway, M., and M. Scheiman. The efficacy of vision therapy for convergence excess. *Journal of the American Optometric Association* 68 (1997): 81–85.

Garzia, R. P. The efficacy of visual training in amblyopia: A literature review. *American Journal of Optometry and Physiological Optics* 64 (1987): 393–404.

Garzia, R., and J. Richman. Accommodative facility: A study of young adults. *Journal of the American Optometric Association* 53 (1982): 821–24.

Goldrich, S. G. Optometric therapy of divergence excess strabismus. *American Journal of Optometry and Physiological Optics* 57 (1980): 7–14.

———. Oculomotor biofeedback and intermittent exotropia. *American Journal of Optometry and Physiological Optics* 59 (1982): 306–17.

Grisham, J. D., M. C. Bowman, A. Owyang, and C. L. Chan. Vergence orthoptics: Validity and persistence of the training effect. *Optometry and Vision Science* 68 (1991): 441–51.

Halliwell, J. W., and H. A. Solan. The effects of a supplemental perceptual training program on reading achievement. *Exceptional Children* (1972): 613–21.

Haynes, H. M., and L. G. McWilliams. Effects of training on near-far response time as measured by the distance rock test. *Journal of the American Optometric Association* 50 (1979): 715–18.

Hennessey, D., and R. Iosue. Relation of symptoms to accommodative infacility of school-aged children. *American Journal of Optometry and Physiological Optics* 61 (1984): 177.

Hoffman, L. H. Incidence of vision difficulties in children with learning disabilities. *Journal of the American Optometric Association* 51 (1980): 447–51.

———. The effect of accommodative deficiencies on the development level of perceptual skills. *American Journal of Optometry and Physiological Optics* 59 (1982): 254–62.

Hoffman, L. H., and A. H. Cohen. Effectiveness of non-strabismic optometric vision training in a private practice. *American Journal of Optometry and Archives of the American Academy of Optometry* 47 (1973): 813–16.

Hung, G. K., K. J. Ciuffreda, and J. L. Semmlow. Static vergence and accommodation: Norms and orthoptic effects. *Documents of Ophthalmology* 62 (1986): 165–79.

Lieberman, S. The prevalence of visual disorders in a school for emotionally disturbed children. *Journal of the American Optometric Association* 56 (1985): 800–805.

Levine, S., et al. Clinical assessment of accommodative facility in symptomatic and asymptomatic individuals. *Journal of the American Optometric Association* 56 (1985): 286–90.

Ludlam, W. M. Orthoptic treatment of strabismus. *American Journal of Optometry and Archives of the American Academy of Optometry* 38 (1961): 369–88.

———. Visual training, the alpha activation cycle and reading. *Journal of the American Optometric Association* 50 (1979): 111–15.

Ludlam, W. M., and B. I. Kleinman. The long-range results of orthoptic treatment of strabismus. *American Journal of Optometry and Archives of the American Academy of Optometry* 42 (1965): 647–84.

Pavlidis, G. T. Eye movements in dyslexia: Their diagnostic significance. *Journal of Learning Disabilities* 18 (1985): 42–50.

Poynter, H. L., et al. Oculomotor functions in reading disability. *American Journal of Optometry and Physiological Optics* 59 (1982): 116–27.

Press, L. J. The interface between ophthalmology and optometric vision therapy. *Binocular Vision and Strabismus Quarterly* 17 (2002): 6–11.

Richman, J. E. Use of a sustained visual attention task to determine children at risk for learning problems. *Journal of the American Optometric Association* 57 (1986): 20–26.

Saladin, J. J., and J. O. Rick. Effect of orthoptic procedures on stereoscopic acuities. *American Journal of Optometry and Physiological Optics* (1982): 718–25.

Seiderman, A. S. Optometric vision therapy—results of a demonstration project with a learning disabled population. *Journal of the American Optometric Association* 51 (1980): 489–93.

Sheedy, J. E., and J. J. Saladin. Association of symptoms with measures of oculomotor deficiencies. *American Journal of Optometry and Physiological Optics* 55 (1978): 670–76.

Solan, H. A., et al. Role of visual attention in cognitive control of oculomotor readiness in students with reading disabilities. *Journal of Learning Disabilities* 34 (2001): 107–18.

Solan, H. A., and A. P. Ficarra. A study of perceptual and verbal skills of disabled readers in grades 4, 5, and 6. *Journal of the American Optometric Association* 61 (1990): 628–34.

Solan, H. A., A. P. Ficarra, J. R. Brannan, and F. Rucker. Eye movement efficiency in normal and reading disabled elementary school children: Effects of varying luminance and wavelength. *Journal of the American Optometric Association* 69 (1998): 455–64.

Suchoff, I. B., and G. T. Petito. The efficacy of visual therapy: Accommodative disorders and non-strabismic anomalies of binocular vision. *Journal of the American Optometric Association* 57 (1986): 119–25.

Vaegan, J. L. Convergence and divergence show large and sustained improvement after short isometric exercise. *American Journal of Optometry and Physiological Optics* 56 (1979): 23–33.

Wittenberg S., F. W. Brock, and W. C. Folsom. Effect of training on stereoscopic acuity. *American Journal of Optometry and Archives of the American Academy of Optometry* 46 (1969): 645–53.

Zellers, J. A., T. L. Alpert, and M. W. Rouse. A review of the literature and a normative study of accommodative facility. *Journal of the American Optometric Association* 55 (1984): 31–37.

Bibliography

Atzmon, D., P. Nemet, A. Ishay, and E. Karni. A randomized prospective masked and matched comparative study of orthoptic treatment versus conventional reading tutoring treatment for reading disabilities in 62 children. *Binocular Vision and Eye Muscle Surgery Quarterly* 8 (1993): 91–106.

Benson, K., and B.K. Hartz. A comparison of observational studies and randomized, controlled trials. *New England Journal of Medicine* 342 (2002): 1878–86.

Cook, David L. *When Your Child Struggles: The Myths of 20/20 Vision* (Atlanta: Invision Press, 1992).

———. Vision therapy and quality of life. *Journal of Optometric Vision Development* 26 (1995). 205–11.

Cooper, Andrew. *Playing in the Zone: Exploring the Spiritual Dimensions of Sports* (Boston: Shambhala, 1998).

Cooper, J., and N. Medow. Intermittent exotropia basic and divergence excess type. *Binocular Vision and Eye Muscle Surgery Quarterly* (1993): 185–216.

Forrest, Elliot B. *Visual Imagery: An Optometric Approach* (Santa Ana, CA: OEP Foundation, 1981).

Fountain, Paul. "Eyes That See." *Flying*, August 1945.

Gallwey, W. Timothy. *The Inner Game of Tennis* (New York: Random House, 1974).

Getman, Gerald N. "The Mileposts to Maturity." *The Optometric Weekly*, April 6, 1972.

Harrington, Anne, ed. *The Placebo Effect: An Interdisciplinary Exploration* (Cambridge, MA: Harvard University Press, 1997).

Hoare, Arthur E. "The Skeffington Saga—Part I: Skeffington, The Man." *Optometric World,* May 1966.

———. "The Skeffington Saga—Part 2: Skeffington, The Mission." *Optometric World,* July 1966.

Kavner, Richard S., and Lorraine Dusky. *Total Vision* (New York: Kavner Books, 1978).

MacDonald, Lawrence W. *Visual Training*. Series 1–2. (Santa Ana, CA: OEP Foundation, 1978–79).

McPhee, John. *A Sense of Where You Are: A Profile of Bill Bradley at Princeton* (New York: Farrar, Straus, and Giroux, 1965).

Revell, M. J. *Strabismus: A History of Orthoptic Techniques* (London: Barrie and Jenkins, 1971).

Romano, P. E. Optometric vision therapy and training for learning disabilities and dyslexia; DVD surgery; curing complications of strabismus surgery. *Binocular Vision and Strabismus Quarterly* 17 (2002): 12–14.

Seiderman, Arthur S., Steven E. Marcus, and David Hapgood. *20/20 Is Not Enough: The New World of Vision* (New York: Fawcett Crest, 1994).

Shepard, Carl F. Optometric indoctrination. *The Optometric Weekly,* June 14, 1951.

Williams, Ted, and John Underwood. *My Turn at Bat* (New York: Fireside, 1988).

Woods, Tiger, with the Editors of *Golf Digest. How I Play Golf* (New York: Warner Books, 2001).

Yandell, John. *Visual Tennis: Using Mental Imagery to Perfect Your Stroke Technique* (Champaign, IL: Human Kinetics, 1999).